de Gruyter Series in Nonlinear Analysis and Applications 6

Jianhong Wu

Introduction to Neural Dynamics and Signal Transmission Delay

Walter de Gruyter · Berlin · New York 2001

Author

Jianhong Wu
Department of Mathematics and Statistics
York University
4700 Keele Street
North York, Ontario M3J 1P3
Canada

Mathematics Subject Classification 2000: 34-02; 34K20, 34K15, 92B20

Keywords: Neural networks, Delay, Dynamics, Associative memory

⊗ Printed on acid-free paper which falls within the guidelines of the ANSI
to ensure permanence and durability.

Library of Congress Cataloging-in-Publication Data

Wu, Jianhong.
Introduction to neural dynamics and signal transmission
delay / by Jianhong Wu.
p.   cm.
Includes bibliographical references and index.
ISBN 3-11-016988-6
1. Neural networks (Neurobiology) − Mathematical models.   I. Title.
QP363.3 .W8   2001
573.8'5−dc21   2001017248

Die Deutsche Bibliothek − Cataloging-in-Publication Data

Wu, Jianhong:
Introduction to neural dynamics and signal transmission delay /
Jianhong Wu. − Berlin ; New York : de Gruyter, 2001
(De Gruyter series in nonlinear analysis and applications ; 6)
ISBN 3-11-016988-6

ISSN 0941-813 X

Printed in Germany.
Typesetting using the authors' TEX files: I. Zimmermann, Freiburg.
Printing and binding: WB-Druck GmbH & Co., Rieden/Allgäu.
Cover design: Thomas Bonnie, Hamburg.

To Ming

Preface

In the design of a neural network, either for biological modeling, cognitive simulation, numerical computation or engineering applications, it is important to describe the dynamics (also known as evolution) of the network. The success in this area in the early 1980's was one of the main sources for the resurgence of interest in neural networks, and the current progress towards understanding neural dynamics has been part of exhaustive efforts to lay down a solid theoretical foundation for this fast growing theory and for the applications of neural networks.

Unfortunately, the highly interdisciplinary nature of the research in neural networks makes it very difficult for a newcomer to enter this important and fascinating area of modern science. The purpose of this book is to give an introduction to the mathematical modeling and analysis of networks of neurons from the viewpoint of dynamical systems. It is hoped that this book will give an introduction to the basic knowledge in neurobiology and physiology which is necessary in order to understand several popular mathematical models of neural networks, as well as to some well-known results and recent developments in the theoretical study of the dynamics of the mathematical models.

This book is written as a textbook for senior undergraduate and graduate students in applied mathematics. The level of exposition assumes a basic knowledge of matrix algebra and ordinary differential equations, and an effort has been made to make the book as self-contained as possible.

Many neural network models were originally proposed to explain observations in neurobiology or physiology. Also, understanding human behavior and brain function is still one of the main motivations for neural network modeling and analysis. It is therefore of prime importance to understand the basic structure of a single neuron as well as a network of neurons and to understand the main mechanisms of the neural signal transmission. This necessary basic knowledge in neuroscience will be collected in Chapter 1.

Chapter 2 will start with the derivation of general models of biological neural networks. In particular, the additive and shunting equations will be presented and a brief description of the popular signal functions will be given. An abstract formulation, which is suitable for both biological and artificial networks and treats a network as a labeled graph, will be provided as well. Two types of network architectures: feedforward and feedback networks will be identified and various connection topologies will be described.

In Chapter 3, several simple networks will be presented that perform some elementary functions such as storing, recalling and recognizing neuron activation patterns. The important on-center off-surround connection topology and its connection to the solution of the noise-saturation dilemma will be discussed. Two important issues related to applications: the choice of signal functions and the determination of the synaptic coupling coefficients will also be addressed.

A central subject of this book is the long-term behaviors of the network. In Chapter 4, the connection between the convergence and the global attractor and the important property of content-addressable memory of many networks will be discussed. The convergence theorem due to Cohen, Grossberg and Hopfield based on LaSalle's Invariance Principle and its various extensions and modifications will be presented. Other techniques for establishing the convergence of almost every trajectory of a given network will also be provided, including the theory of monotone dynamical systems and the combinatorial matrix theory.

A special feature of this book is its emphasis on the effect of signal delays on the long-term performance of the networks under consideration. Such time lags exist due to the finite propagation speeds of neural signals along axons and the finite speeds of the neurotransmitters across synaptic gaps in a biological neural network and due to the finite switching speeds of amplifiers (neurons) in artificial neural networks. In the final chapter, various phenomena associated with the signal delays: delay-induced instability, nonlinear periodic solutions, transient oscillations, phase-locked oscillations and changes of the basins of attraction will be demonstrated.

This book grows from the lecture notes given by the author during a summer graduate course in Neural Dynamics at York University in 1999. I am deeply indebted to all the students form this class. I especially want to thank Yuming Chen who typed the whole text, struggled with evolving versions of it and offered many critical comments and suggestions.

Professor Hugh R. Wilson read several chapters of the manuscript and offered many helpful comments which are greatly appreciated. I thank my editor, Manfred Karbe, for his patience, understanding, encouragement and assistance.

It is also a pleasure to acknowledge the financial support from Natural Sciences and Engineering Research Council of Canada, and from the Network of Centers of Excellence: Mathematics for Information Technology and Complex Systems.

Toronto, February 2001   Jianhong Wu

Contents

Chapter 1

The structure of neural networks

The human central nervous system is composed of a vast number of interconnected cellular units called *nerve cells* or *neurons*. While neurons have a wide variety of shapes, sizes and location, most neurons are of rather uniform structures and neural signals are transmitted on the same basic electrical and chemical principles.

The purpose of this chapter is to give a brief description about the structure of a single neuron and a neural circuit, and about the electrical and chemical mechanism of neural signal transmission. Most materials in this chapter are taken from Harvey [1994] and Purves, Augustine, Fitzpatrick, Katz, LaMantia and McNamara [1997], related references are Grossberg [1982, 1988], Haykin [1994], Levine [1991], Müller and Reinhardt [1991].

1.1 The structure of a single neuron

Figure 1.1.1 shows a prototypical neuron. The central part is called the *cell body* or *soma* which contains nucleus and other organelles that are essential to the function of all cells. From the cell body project many root-like extensions (the *dendrites*) as well as a single tubular fiber (the *axon*) which ramifies at its end into a number of small branches.

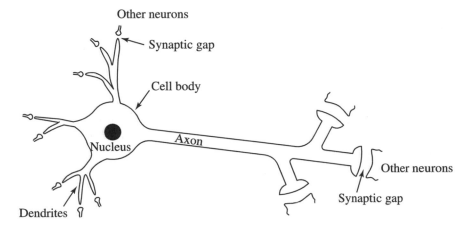

Figure 1.1.1. Schematic structure of a typical neuron.

Dendrites are branch-like protrusions from the neuron cell body. A typical cell has many dendrites that are highly branched. The receiving zones of signals (or impulses or information), called *synapses*, are on the cell body and dendrites. Some neurons have *spines* on the dendrites, thus creating specialized receiving sites. Since a fundamental purpose of neurons is to integrate information from other neurons, the number of inputs received by each neuron is an important determinant of neural function.

The axon is a long fiber-like extension of the cell body. The axon's purpose is signal conduction, i.e., transmission of the impulses to other neurons or muscle fibers. The axonal mechanism that carries signals over the axon is called the *action potential*, a self-regenerating electrical wave that propagates from its point of initiation at the cell body (called the axon *hillock*) to the terminals of the axon. Axon terminals are highly specialized to convey the signal to target neurons. These terminal specializations are called *synaptic endings*, and the contacts they make with the target neurons are called *chemical synapses*. Each synaptic ending contains secretory organelles called *synaptic vesicles*.

The axon fans out to other neurons. The fan-out is typically $1 : 10,000$ and more. The same signal encoded by action potentials propagates along each branch with varying time delays. As mentioned above, each branch end has a synapse which is typically of a bulb-like structure. The synapses are on the cell bodies, dendrites and spines of target neurons. Between the synapses and the target neurons is a narrow gap (called *synaptic gap*), typically 20 nanometers wide. Structures are spoken of in relation to the synapses as *presynaptic* and *postsynaptic*. Special molecules called *neurotransmitters* are released from synaptic vesicles and cross the synaptic gap to receptor sites on the target neuron. The transmission of signals between synapses and target neurons is the flow of neurotransmitter molecules.

1.2   Transmission of neural signals

Neuron signals are transmitted either electrically or chemically. Electrical transmission prevails in the interior of a neuron, whereas chemical mechanisms operate at the synapses. These two types of signaling mechanisms are the basis for all the information-processing capabilities of the brain.

An electrochemical mechanism produces and propagates signals along the axon. In the state of inactivity (*equilibrium*) the interior (*cytoplasm*) of the neuron and the axon is negatively charged relative to the surrounding extra-cellular fluid due to the differences in the concentrations of specific ions across the neuron membrane. For example, in an exceptionally large neuron found in the nervous system of the squid, ion concentrations of potassium (K^+), sodium (Na^+), chloride (Cl^-) and calcium (Ca^{2+}) are 400, 50, 40–150 and 0.0001 mM inside the neuron, and 20, 440, 560 and 10 mM outside the neuron. Such measurements are the basis for the claim that there is much more K^+ inside the neuron than out, and much more Na^+ outside than in. Similar concentration gradients occur in the neurons of most animals, including humans. The equilibrium ion concentration gradients, called *the resting potential*, of about −70 mV (typically −50 to −80 mV, depending on the type of neurons being examined) is caused by a biological ion pump powered by mitochondria (structures in the neuron which use and transfer the energy produced by burning fuel (sugar) and molecular oxygen).

The excess of positive and negative charges generates an electric field across the neuron surface and the plasma membrane of the neuron holds the charges apart. See Figure 1.2.1.

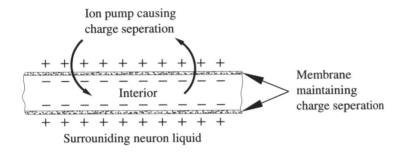

Figure 1.2.1. Biological ion pump powered by mitochondria.

The membrane is selective to diffusion of particular molecules, and its diffusion selectivity varies with time and length along the axon. This selective permeability of membranes is due largely to *ion channels* which allow only certain kinds of ions to cross the membrane in the direction of their concentration and electro-chemical gradients. Thus, channels and pumps basically work against each other.

The axon signal pulses are the fundamental electrical signals of neurons. They are called *action potentials* (or *spikes* or *impulses*) and are described electrically by current-voltage characteristics. Injecting a current pulse into the axon causes the potential across the membrane to vary. As mentioned above, equilibrium potential of the axon is about −70 mV. When injected an inhibitory (*hyperpolarization*) or excitatory (*depolarization*) signal pulse, the response is a RC-type (exponential) followed by relaxation. When the injected current causes the voltage to exceed the threshold (typically −55 mV), a special electrochemical process described below generates a rapid increase in the potential and this injected current then produces a single pulse with a peak-to-peak amplitude of about 100–120 mV and a duration of 0.5–1 ms. This generated pulse propagates without weakening along all branches of the axon. A complex sequence involving the ions maintains the pulse shape and strength. A brief description of the production and propagation of a pulse by a step-by-step electrochemical process is as follows (see Figure 1.2.2):

1. Chemical signals at excitatory synapses inject current in the axon;
2. Rising positive charges in the cell leads to the sudden loss of impermeability

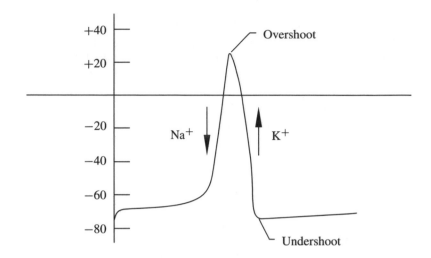

Figure 1.2.2. Production and propagation of a pulse.

against Na^+ ions of the membrane and trigger Na^+ gate along the membrane to open (threshold requires a depolarization of about 10 mV from rest);

3. Na^+ ions enter the interior until the Na^+ equilibrium potential is almost reached. As the Na^+ equilibrium potential is typically near +50 mV, the interior of the neuron even acquires a positive potential against its surroundings (overshooting);

4. The positive interior causes adjacent Na^+ gates to open so that an impulse propagates down the axon;

5. The Na^+ gates inactivate[1] and the outflow of K^+ from the interior causes the axon to hyperpolarize.

As a result of Na^+ channel inactivation and K^+ outflow, the frequency of impulse generation is limited. For one or a few milliseconds after an impulse, no additional impulses can be generated. This is called the *absolute refractory period*. For several more milliseconds afterwards, there is a *relative refractory period* in which an action potential can be initiated, but the strength of current required to initiate an impulse is larger than normal (Levine [1991]).

The pulse normally propagates from the cell body to the terminals of the neuron. The speed of propagation of the pulse signal along the axon varies greatly. In the cells of the human brain the signal travels with a velocity of about 0.2–2 m/s. For other

---

[1]In fact, there are two factors to cause the membrane potential to return to the resting potential after an action potential. The most important is that after a Na^+ channel opens, it closes spontaneously, even though the membrane potential is still depolarized. This process, called Na^+ inactivation, is about ten times slower than Na^+ channel opening. In addition to Na^+ inactivation, depolarization opens voltage-gated K^+ channels, increasing the K^+ permeability above that in the resting cell. Opening voltage-gated K^+ channels favors more rapid repolarization. Voltage-gated K^+ channels activate at about the same rate that Na^+ channels inactivate. See Raven and Johnson [1995].

cells, the transmission velocity can be up to 100 m/s due to the so-called *saltatory conduction* between *Ranvier nodes*. This is because many axons are covered by an electrically insulating layer known as the *myelin sheath* which is interrupted from time to time at the so-called Ranvier nodes, and the signal jumps along the axon from one Ranvier node to the next.

The above discussion describes the production and propagation of a single pulse. It should be emphasized that the amplitude of the single pulse is independent of the magnitude of the current injected into the axon which generates the pulse. In other words, larger injected currents do not elicit larger action potentials. The action potentials of a given neuron are therefore said to be all-or-none. On the other hand, if the amplitude or duration of the injected current is increasing significantly, multiple action potentials occur. Therefore, one pulse does not carry information, and pulses are produced randomly due to the massive receiving sites of a single neuron. The intensity of a pulse train is not coded in the amplitude, rather it is coded in the frequency of succession of the invariant pulses, which can range from about 1 to 1000 per second. The interval between two spikes can take any value (larger than the absolute refractory period), and the combination of analog and digital signal processing is utilized to obtain security, quality and simplicity of information processing.

The pulse signal traveling along the axon comes to a halt at the synapse due to the synaptic gap. The signal is transferred to the target neuron across the synaptic gap mainly via a special chemical mechanism, called *synaptic transmission*. In synaptic transmission, when a pulse train signal arrives at the presynaptic site, special substances called *neurotransmitters* are released. The neurotransmitter molecules diffuse across the synaptic gap, as shown in Figure 1.2.3, reaching the postsynaptic neuron (or

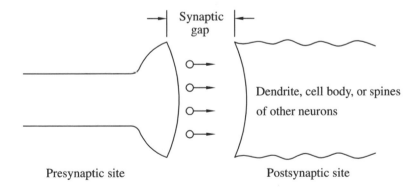

Figure 1.2.3. Signal transmission at the synaptic gap.

muscle fiber) within about 0.5 ms. Upon their arrival at special receptors, these substances modify the permeability of the postsynaptic membrane for certain ions. These ions then flow in or out of the neurons, causing a hyperpolarization or depolarization of the local postsynaptic potential. If the induced polarization potential is positive (resp. negative), then the synapse is termed *excitatory* (resp. *inhibitory*), because the influence of the synapse tends to activate (resp. to inhibit) the postsynaptic neuron. The differences in excitation and inhibition are primarily membrane permeability differences. Table 1.2.4 gives a comparison of the sequences for synaptic excitation and inhibition.

Excitation	Inhibition
1. Presynaptic sites release excitatory neurotransmitter (such as Acetylcholine (ACh^+) or glutamate).	1. Presynaptic sites release inhibitory neurotransmitter (such as γ-aminobutyric acid (GABA)).
2. The neurotransmitter diffuses across the synaptic gap to the postsynaptic membrane.	2. The neurotransmitter diffuses across the synaptic gap to the postsynaptic membrane.
3. The postsynaptic membrane permeability to Na^+ and K^+ greatly increases.	3. The postsynaptic membrane permeability to Cl^- and K^+ greatly increases.
4. Na^+ and K^+ ion currents across the membrane drive the potential of the target neuron to the $-20\,mV$-to-$0\,mV$ range (usually near $0\,mV$).	4. Cl^- and K^+ ion currents across the membrane drive the potential of the target neuron below $-70\,mV$.

Table 1.2.4. Sequences for synaptic excitation and inhibition.

Note that inhibitory synapses sometimes terminate at the presynaptic sites of other axons, inhibiting their ability to send neurotransmitters across the synaptic gap. This *presynaptic inhibition* is helpful when many pathways converge because the system can selectively suppress inputs.

The polarization potential caused by a single synapse might or might not be large enough to depolarize the postsynaptic neuron to its firing threshold. In fact, the postsynaptic neuron typically has many dendrites receiving synapses from thousands of different presynaptic cells (neurons). Hence its firing depends on the sum of depolarizing effects from these different dendrites. These effects decay with a characteristic time of 5–10 ms, but if signals arrive at the same synapse over such a period, then excitatory effects accumulate. A high rate of repetition of firing of a neuron therefore expresses a large intensity of the signal. When the total magnitude of the depolarization potential in the cell body exceeds the critical threshold (about 10 mV), the neuron fires.

The neurotransmitter is randomly emitted at every synaptic ending in quanta of a few 1000 molecules at a low rate. The rate of release is increasing enormously upon arrival of an impulse, a single action potential carrying the emission of 100–300 quanta within a very short time.

1.3   Neural circuits, CNS and ANN

The complexity of the human nervous system and its subsystems rests on not only the complicated structures of single nerve cells and the complicated mechanisms of nerve signal transmission, but also the vast number of neurons and their mutual connections.

Neurons do not function in isolation. They are organized into ensembles called *circuits* that process specific kind of information. Although the arrangement of neural circuitry is extremely varied, the information of neural circuits is typically carried out in a dense matrix of axons, dendrites and their connections. Processing circuits are combined in turn into *systems* that serve broad functions such as the visual system and the auditory system.

The nervous system is structurally divided into central and peripheral components. The *central nervous system* (CNS) comprises the brain (cerebrum, cerebellum and the brainstem) and the spinal cord. The *peripheral nervous system* includes sensory neurons, which connect the brain and the spinal cord to sensory receptors, as well as motor neurons, which connect brain and the spinal cord to muscles and glands.

Connectivity of neurons is essential for the brain to perform complex tasks. In the human cortex, every neuron is estimated to receive converging input on the average from about 10^4 synapses. On the other hand, each neuron feeds its output into many hundreds of other neurons, often through a large number of synapses touching a single nerve cell. It is estimated that there must be in the order of 10^{11} neurons in the human cortex, and 10^{15} synapses.

Plasticity is another essential feature of neural networks. There is a great deal of evidence that the strength of a synaptic coupling between two given neurons is not fixed once and for all. As originally postulated by Hebb [1949], the strength of a synaptic connection can be adjusted if its level of activity changes. An active synapse, which repeatedly triggers the activation of its postsynaptic neuron, will grow in strength, while others will gradually weaken. This plasticity permits the modification of synaptic coupling and connection topology, which is important in the network's ability of learning from, and adapting to its environments.

The brain's capability of organizing neurons to perform certain complex tasks comes from its massively parallel distributed structure and its plasticity. This complex and plastic parallel computer has the capability of organizing neurons so as to perform certain computations such as pattern recognition and associative memory many times faster than the fastest digital computer. It is therefore of great importance and interest to understand how the biological neural networks perform these computations and how to build a biologically motivated machine to perform these computations.

An *artificial neural network* (ANN) is a machine that is designed to model the way in which the brain performs a particular task or function of interest. Such a network is usually implemented using electronic components or simulated in software on a digital computer. In the last two decades, neurobiology and experimental psychology have been rapidly developed and great progress has been made in the expansion of cognitive or adaptive capabilities in industrial applications of computers. Many designers of intelligence machines have borrowed ideas from experimental results in the brain's analog response, and machines have been built like simulated brain regions/functions, with *nodes* corresponding to neurons or populations of neurons, and connections between the nodes corresponding to the synapses of neurons. As these fields develop further and interact, it is natural to look for common organizing principles, quantitative foundations and mathematical modeling.

We conclude this chapter with the following definition, from the DARPA study [1988], which seems to be suitable for both biological and artificial neural networks:

> A neural network is a system composed of many simple processing elements operating in parallel whose function is determined by network structure, connection strengths, and the processing performed at computing elements, …. Neural network architectures are inspired by the architecture of biological nervous systems operating in parallel to obtain high computation rates.

Note that, in the above definition and throughout the remaining part of this book, we will alternatively call the functional units in neural networks "nodes", "units", "cells", "neurons" (and even "populations" if a group of neurons is represented by a node, see Section 2.1).

Chapter 2

Dynamic models of networks

In this chapter, we start with a general model to describe the dynamics of a biological network of neurons. As often in the development of science, the derivation of the model equations based on well-accepted laws and principles is the starting point to understand human brain-mind functioning and to design a machine (artificial neural network) to mimic the way in which the brain performs a particular task of interest. We will also give an abstract formulation of a network as a labeled graph and discuss two important network architectures.

2.1   Biological models

Although it is well known that a neuron transmits voltage spikes along its axon, many studies show that the effects on the receiving neurons can be usefully summarized by voltage potentials in the neuron interior that vary slowly and continuously during the time scale of a single spike. That is, the time scale of the interior neuron potential is long compared to the spike duration. The biological model of neural networks introduced below, following the presentation of Harvey [1994], reflects this viewpoint.

Assume that the network, shown in Figure 2.1.1, consists of n neurons, denoted by $v_1, \ldots, v_n$. We introduce a variable x_i to describe the neuron's state and a variable Z_{ij} to describe the coupling between two neurons v_i and v_j. More precisely, let

$$x_i(t) = \text{deviation of the } i\text{th neuron's potential from its equilibrium.}$$

This variable describes the activation level of the ith neuron. It is called the action potential, or the *short-term memory* (STM) trace.

The variable Z_{ij} associated with v_i's interaction with another neuron v_j is defined as

$$Z_{ij} = \text{neurotransmitter average release rate per unit axon signal frequency.}$$

This is called the *synaptic coupling coefficient* (or *weight*) or the *long-term memory* (LTM) trace.

Let us first derive the so-called additive STM equation. Assume a change in neuron potential from equilibrium ($-70\,\text{mV}$) occurs. In general, the change is caused by

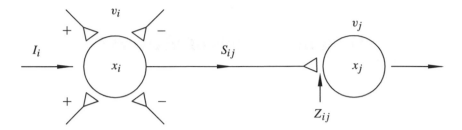

Figure 2.1.1. Schematic diagram of a neuron and a network: Neuron v_i with potential x_i relative to equilibrium sends a signal S_{ij} along the axon to a target neuron v_j. The signal affects the target neuron with a coupling strength Z_{ij}. The signal may be excitatory ($Z_{ij} > 0$) or inhibitory ($Z_{ij} < 0$). Other inputs I_i to the neuron model external stimuli.

internal and external processes. Thus, we have the following form for the STM trace:

$$\frac{dx_i}{dt} = \left(\frac{dx_i}{dt}\right)_{\text{internal}} + \left(\frac{dx_i}{dt}\right)_{\text{external}}. \qquad (2.1.1)$$

Assume inputs from other neurons and stimuli are additive. Then

$$\frac{dx_i}{dt} = \left(\frac{dx_i}{dt}\right)_{\text{internal}} + \left(\frac{dx_i}{dt}\right)_{\text{excitatory}} - \left(\frac{dx_i}{dt}\right)_{\text{inhibitory}} + \left(\frac{dx_i}{dt}\right)_{\text{stimuli}}. \qquad (2.1.2)$$

Assuming further the internal neuron processes are stable (that is, the neuron's potential decays exponentially to its equilibrium without external processes), we have

$$\left(\frac{dx_i}{dt}\right)_{\text{internal}} = -A_i(x_i)x_i, \quad A_i(x_i) > 0. \qquad (2.1.3)$$

In most models discussed in this book and in much of the literature, A_i is assumed to be a constant. But, in general, A_i can vary as a function of x_i. See the example provided at the end of this section.

Assume additive synaptic excitation is proportional to the pulse train frequency. Then

$$\left(\frac{dx_i}{dt}\right)_{\text{excitatory}} = \sum_{\substack{k=1 \\ k \neq i}}^{n} S_{ki} Z_{ki}, \qquad (2.1.4)$$

where S_{ki} is the average frequency of signal evaluated at v_i in the axon from v_k to v_i. This is called the *signal function*. In general, S_{ki} depends on the propagation time delay τ_{ki} from v_k to v_i and the threshold Γ_k for firing of v_k in the following manner

$$S_{ki}(t) = f_k(x_k(t - \tau_{ki}) - \Gamma_k) \qquad (2.1.5)$$

for a given nonnegative function $f_k : \mathbb{R} \to [0, \infty)$. Commonly used forms of the signal function will be described in next section. Here and in what follows, the propagation *time delay* τ_{ki} is the time needed for the signal sent by v_k to reach the receiving site of neuron v_i, and the *threshold* Γ_k is the depolarization at which the neuron v_k fires (*i.e.*, transmits an impulse).

Assume hardwiring of the inhibitory inputs from other neurons, that is, their coupling strength is constant. Then

$$\left(\frac{dx_i}{dt} \right)_{\text{inhibitory}} = \sum_{\substack{k=1 \\ k \neq i}}^{n} C_{ki} \tag{2.1.6}$$

with

$$C_{ki}(t) = c_{ki} f_k(x_k(t - \tau_{ki}) - \Gamma_k), \tag{2.1.7}$$

where $c_{ki} \geq 0$ are constants.

We now turn to the LTM trace. Assume the excitatory coupling strength varies with time. A common model based on Hebb's law is

$$\frac{dZ_{ij}}{dt} = -B_{ij}(Z_{ij})Z_{ij} + P_{ij}[x_j]^+, \qquad B_{ij}(z_{ij}) > 0, \tag{2.1.8}$$

where

$$P_{ij}(t) = d_{ij} f_i(x_i(t - \tau_{ij}) - \Gamma_i), \tag{2.1.9}$$

here $d_{ij} \geq 0$ are constants and $[z]^+ = z$ if $z \geq 0$, and $[z]^+ = 0$ if $z < 0$. The term $P_{ij}[x_j]^+$ shows that to increase Z_{ij}, v_i must send a signal P_{ij} to v_j and, at the same time, v_j must be activated ($x_j > 0$).

Putting the above formulations and the external stimuli together, we then get the following additive STM and passive decay LTM system, for $1 \leq i \leq n, 1 \leq j \leq n$,

$$\begin{cases} \dot{x}_i(t) = -A_i(x_i(t))x_i(t) + \sum_{\substack{k=1 \\ k \neq i}}^{n} f_k(x_k(t - \tau_{ki}) - \Gamma_k)Z_{ki}(t) \\ \qquad - \sum_{\substack{k=1 \\ k \neq i}}^{n} c_{ki} f_k(x_k(t - \tau_{ki}) - \Gamma_k) + I_i(t), \\ \dot{Z}_{ij}(t) = -B_{ij}(Z_{ij}(t))Z_{ij}(t) + d_{ij} f_i(x_i(t - \tau_{ij}) - \Gamma_i)[x_j(t)]^+. \end{cases} \tag{2.1.10}$$

Note that we use $\dot{x}_i(t)$ for $\frac{d}{dt}x_i(t)$ and $\dot{Z}_{ij}(t)$ for $\frac{d}{dt}Z_{ij}(t)$.

We next derive the STM shunting equations. To motivate the shunting equation, we start with the so-called membrane equation on which cellular neurophysiology is based. The membrane equation, appeared in the Hodgkin–Huxley model (see, for example, Hodgkin and Huxley [1952]), describes the voltage $V(t)$ of a neuron by

$$C \frac{dV}{dt} = (V^+ - V)g^+ + (V^- - V)g^- + (V^P - V)g^P \tag{2.1.11}$$

shown in Figure 2.1.2, where C is the capacitance, V^+, V^- and V^P are constant

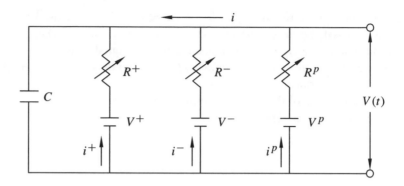

Figure 2.1.2. Circuit analogy for a neuron membrane. V^+, V^-, V^P are the excitatory, inhibitory and passive saturation points, respectively, inside a neuron. The voltages act like batteries in an electrical circuit. They produce a fluctuating output voltage $V(t)$ representing the action potential inside the neuron.

excitatory, inhibitory and passive saturation points, respectively, and g^+, g^- and g^P are excitatory, inhibitory and passive conductance, respectively. Often V^+ represents the saturation point of a Na^+ channel and V^- represents the saturation point of a K^+ channel.

At each neuron v_i, we let

$$\begin{cases} V = x_i, \\ V^+ = B_i, \\ V^- = -D_i, \\ V^P = 0, \\ g^+ = I_i + \sum_{k \neq i} S_{ki}^{(+)} Z_{ki}^{(+)}, \\ g^- = J_i + \sum_{l \neq i} S_{li}^{(-)} Z_{li}^{(-)}, \\ g^P = A, \\ C = 1 \text{ (rescale time).} \end{cases} \qquad (2.1.12)$$

Then we obtain the following shunting STM equation

$$\begin{aligned} \dot{x}_i = &-A_i x_i + (B_i - x_i)\Big[\sum_{k \neq i} S_{ki}^{(+)} Z_{ki}^{(+)} + I_i\Big] \\ &- (x_i + D_i)\Big[\sum_{l \neq i} S_{li}^- Z_{li}^{(-)} + J_i\Big], \quad 1 \leq i \leq n, \end{aligned} \qquad (2.1.13)$$

where

$$\begin{aligned} S_{ij}^{(+)}(t) &= f_i(x_i(t - \tau_{ij}) - \Gamma_i), \\ S_{ij}^{(-)}(t) &= f_i(x_i(t - \tau_{ij}) - \Gamma_i), \quad 1 \leq j \neq i \leq n, \end{aligned}$$

and $Z_{ki}^{(+)}$ and $Z_{ki}^{(-)}$ are the corresponding LTM traces of the respective excitatory and inhibitory synaptic coupling. In the above shunting equations, B_i and $-D_i$ are the maximal and minimal action potentials of neuron v_i. As we will see in subsequent chapters, in networks modeled by shunting equations, there exists some global control on the dynamics of the network which prevents too high or too low activity. Another derivation of shunting equations based on the mass action law will be given later.

Finally, we point out that it is often convenient to use a single node to represent a pool of interacting neurons of similar structures. Figure 2.1.3 shows a typical situation

Interacting neurons

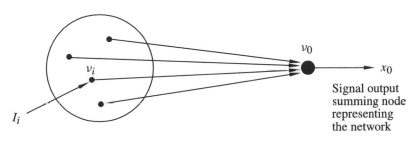

Figure 2.1.3. If the time scales of the input and neuron dynamics are properly chosen, a single neuron v_0 may represent a group of neurons. In this interpretation, the activation level of v_0 is proportional to the fraction of excited neurons in the group.

consisting of a group of neurons and one output representing node. For each neuron in the group, we have

$$\dot{x}_i = -A_i x_i + \sum_{k\neq i}^{n} S_{ki} Z_{ki} - \sum_{l\neq i}^{n} C_{li} + I_i, \quad 1 \leq i \leq n, \qquad (2.1.14)$$

and

$$\dot{Z}_{ij} = -B_{ij} Z_{ij} + P_{ij}[x_j]^+, \quad 1 \leq i, j \leq n, \qquad (2.1.15)$$

where the summation indices reflect the interconnections. For the output node, we have

$$\dot{x}_0 = -A_0 x_0 + \sum_{i=1}^{n} S_{i0} Z_{i0} \qquad (2.1.16)$$

and

$$\dot{Z}_{i0} = -B_{i0} Z_{i0} + P_{i0}[x_0]^+, \quad 1 \leq i \leq n. \qquad (2.1.17)$$

Assume the following holds:

(H1) $Z_{i0} \approx$ constant Z_0 (slowly varying) for $1 \leq i \leq n$.

(H2) $S_{i0} \approx s_\infty(x_i)$ (short delays, low thresholds and step signal function) for $1 \leq i \leq n$, where $s_\infty(x_i) = 1$ if $x_i > 0$ and $s_\infty(x_i) = 0$ if $x_i \leq 0$.

Then (2.1.16) becomes

$$\dot{x}_0 = -A_0 x_0 + \sum_{i=1}^{n} s_\infty(x_i) Z_0.$$

Consequently, if A_0 is large and the time scales for integrating inputs in (2.1.14) are short, then x_0 is proportional to the number of excited neurons in the pool.

This discussion leads to the following formulation for networks of groups of neurons: Consider a network consisting of $v_1, \ldots, v_n$ groups of interacting neurons. Let x_i be defined as follows:

$$x_i = \text{ action potentials of neurons in group } v_i.$$

Assume v_i has B_i excitable sites. In a binary model (that is, neurons are on or off), we can then regard x_i as the number of sites excited and $B_i - x_i$ the number of sites unexcited. However, due to the average random effects over short time intervals in the large number of neurons in group v_i, we may regard x_i as the (continuous) potentials of neurons in v_i. Assuming also x_i decays exponentially to an equilibrium point (arbitrarily set equal to zero), then the mass action leads to the shunting equation

$$\dot{x}_i = -A_i x_i + (B_i - x_i)\Big[\sum_{k \neq i} S_{ki}^+ Z_{ki}^{(+)} + I_i\Big] - (x_i + D_i)\Big[\sum_{l \neq i} S_{li}^{(-)} Z_{li}^{(-)} + J_i\Big]. \quad (2.1.18)$$

Thus, the state equations for a single neuron have the same form as that for a group of interacting neurons, where x_i is the excited states in the group and thus A_i actually depends on x_i.

Throughout the remaining part of this book, we shall use x_i to denote the activation level of a neuron or the excited states in a group, depending on the context and applications. Consequently, it is more appropriate to regard a neuron as a functional unit/node. In general, we will regard the functional units/nodes as collections of neurons sharing some common response properties. This idea has a particular physiology basis in the organization of the visual cortex (Hubel and Wiesel [1962, 1965]) and somatosensory cortex (Mountcastle [1957]) into columns of cells with the same preferred stimuli. Moreover, columns that are close together also tend to have preferred stimuli that are close together. Existence of such columnar organization has also appeared in multimodality association areas of the cortex (Rosenkilde, Bauer and Fuster [1981], Fuster, Bauer and Jerrey [1982], Goldman-Rakic [1984]).

2.2   Signal functions

Recall that

$$S_{ki} = f_k(x_k(t - \tau_{ki}) - \Gamma_k)$$

is the average frequency of signal evaluated at v_i in the axon from v_k to v_i. Typical functions of f_k include

- step function,

- piecewise linear function,

- sigmoid function.

The step function is defined as

$$f(v) = \begin{cases} 1 & \text{if } v \geq 0, \\ 0 & \text{if } v < 0, \end{cases}$$

shown in Figure 2.2.1 (left). Models involving such a signal function are referred to as the McCulloch–Pitts models, in recognition of the pioneering work of McCulloch and Pitts [1943] (the function describes the all-or-none property of a neuron in the McCulloch–Pitts model).

Figure 2.2.1. A step signal function (left) and a piecewise linear signal function (right).

A piecewise linear function is given by

$$f(v) = \begin{cases} 0 & \text{if } v \leq 0, \\ \beta v & \text{if } 0 < v < \frac{1}{\beta}, \\ 1 & \text{if } v \geq \frac{1}{\beta}, \end{cases}$$

as shown in Figure 2.2.1 (right). This describes the nonlinear off-on characteristic of neurons. β is called the *neural gain*. Note that a piecewise linear function reduces to the step function if β is infinity. Such a function has been widely used in *cellular neural network* models. See, for example, Chua and Yang [1988a, 1988b] and Roska and Vandewalle [1993].

The sigmoid function is by far the most common form of a signal function. It is defined as a strictly increasing smooth bounded function satisfying certain concavity and asymptotic properties. An example of a sigmoid function is the *logistic function* given by

$$f(v) = \frac{1}{1 + e^{-4\beta v}}, \quad v \in \mathbb{R},$$

shown in Figure 2.2.2, where $\beta = f'(0) > 0$ is the neuron gain. Other examples
include inverse tangent function and hyperbolic tangent function.

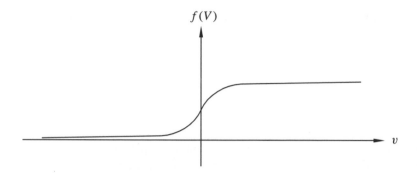

Figure 2.2.2. A sigmoid function $f(v) = \frac{1}{1+e^{-4\beta v}}$.

As $\beta \to \infty$, the sigmoid function becomes the step function. Whereas a step
function assumes the values of 0 or 1, a sigmoid function assumes a continuous
range of values in $(0, 1)$. The smoothness of the sigmoid function is important: it
allows analog signal processing and it makes many mathematical theories applicable.
Another important feature of the bounded sigmoid function is to limit the magnitude
of a nerval signal's impact to its receiving neurons. It was noted, see Levine [1991],
that if the firing threshold of an all-or-none neuron is described by a random variable
with a Gaussian (normal) distribution, then the expected value of its output signal
is a sigmoid function of activity. For this and other reasons, sigmoids have become
increasingly popular in recent neural network models. Also, there has been some
physiological verification of sigmoid signal functions at the neuron level, see Rall
[1955] and Kernell [1965].

Other signal functions are also possible. As will be seen in subsequent chapters,
many global neural network properties are not sensitive to the choice of particular
signal functions, but some are. Choice of a signal function also depends on the
applications considered.

Signal functions are also referred to as *activation functions, amplification functions,
input-output functions* etc in the literature. Such a signal function gives the important
nonlinearity of a neural network.

2.3   General models and network architectures

With the neurobiological analog as the source of inspiration, the theory and applications of ANNs have been advanced rapidly. Recall that by an artificial neural network
we mean a machine that is designed to model the way in which a biological neural
network performs a certain task.

In biological and artificial neural networks, we can regard a neuron as an informa-tion-processing unit. Based on our discussions, for a network of n neurons $v_1, \ldots, v_n$ shown in Figure 2.3.1, we may identify five basic elements associated with neuron v_i (Haykin [1994] and Müller and Reinhardt [1991]):

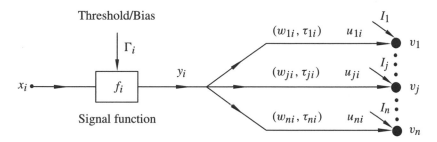

Figure 2.3.1. A general model of five basic elements: activity variable, signal function, threshold/bias, synaptic weight and propagation delay, and external inputs.

(a) a variable x_i describing the *activity* of v_i;

(b) a *signal function* f_i transforming x_i into an *output variable* y_i;

(c) an externally applied *threshold* Γ_i that has the effect of lowering the input of the signal function, or an externally applied *bias* that has the opposite effect of a threshold;

(d) a set of *synapses* connected to other neurons v_j ($1 \leq j \leq n$), each of which is characterized by a *weight* or *strength* w_{ji} and a propagation delay τ_{ji} (the prop-agation delay is always positive, the weight is positive/negative if the associated synapse is excitatory/inhibitory);

(e) an externally applied input I_i.

Viewing the above figure dynamically, we have

$$y_i(t) = f_i(x_i(t) - \Gamma_i),$$
$$u_{ji}(t) = w_{ji} y_i(t - \tau_{ji})$$

and thus the input arising from the signal x_i through the synapse (w_{ji}, τ_{ji}) to the neuron v_j is given by

$$u_{ji}(t) = w_{ji} f_i(x_i(t - \tau_{ji}) - \Gamma_i).$$

We then can define a *neural network* as a *labeled graph* (see Figure 2.3.2) which has the following properties:

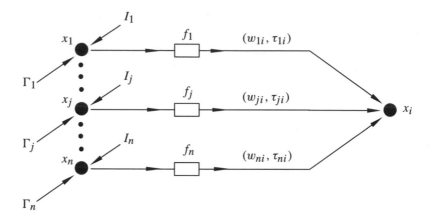

Figure 2.3.2. A network regarded as a labeled graph.

(1) an activity variable x_i is associated with each node i;

(2) a real-valued weight w_{ij} and a nonnegative delay τ_{ij} are associated with link (ji) (called synapse) from node j to node i;

(3) a real-valued threshold/bias Γ_i is associated with node i;

(4) a real-valued function input I_i of time t is associated with node i;

(5) a signal function f_i is defined for node i which determines the output of the node as a function of its threshold/bias and its activity x_i;

(6) the net input to node i is given by the weighted sum $\sum_{j=1}^{n} w_{ij} f_j (x_j(t-\tau_{ij})-\Gamma_j)$ of the outputs of all other nodes connected to i, with corresponding delay.[1]

Recall that nodes are called neurons in the standard terminology. Nodes without links towards them are called *input neurons* and nodes with no link leading away from them are called *output neurons*.

   In general, we may identify two types of network architectures: *feedforward networks* and *feedback networks*. Figure 2.3.3 shows a feedforward network which consists of one layer of 3 neurons (called *input layer*) that projects onto another layer of 2 neurons (called *output layer*), but not vice versa.

   Such a network is called a *single-layer network* (the layer of input neurons is not counted since no information processing is performed here). A more complicated feedforward network is given in Figure 2.3.4, which consists of a layer of 4 input

---

[1]There is another method to calculate the netinput to node i, which is given by $f_i(\sum_{j=1}^{n} w_{ij} x_j (t - \tau_{ij}) - \Gamma_j)$. Models based on such a calculation are usually refereed as Wilson–Cowan equations. See Wilson and Cowan [1972], Grossberg [1972] and Pineda [1987].

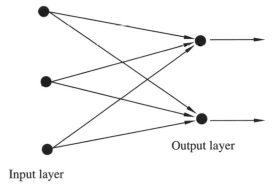

Output layer

Input layer

Figure 2.3.3. A feedforward network consisting of 3 input neurons and 2 output neurons.

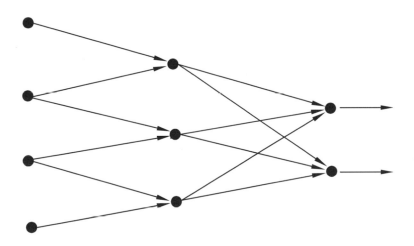

Figure 2.3.4. A network consisting of 3 layers: the input layer, the hidden layer and the output layer.

neurons, a layer of so-called *hidden neurons* and a layer of 2 output neurons. The function of the hidden neurons is to intervene between the external input and the network output. By adding one or more hidden layers, the network is able to extract higher-order statistics, for the network acquires a global perspective despite its local connectivity by virtue of the extra set of synaptic connections and the extra dimension of neural interactions (Churchland and Sejnowski [1992]). A feedforward network is a special example of the so-called cascades of neurons.

In a general multilayer feedforward network, the neurons in the input layer of the network supply respective elements of the activation pattern (input vector), which constitute the input signals applied to the neurons in the second layer (i.e., the first hidden layer). The output signals of the second layer are used as inputs to the third layer, and so on for the rest of the network.

Typically, the neurons in each layer of the network have as their inputs the output signals of the preceding layer only. The set of output signals of the neurons in the output (final) layer of the network constitutes the overall response of the network to the activation pattern supplied by the neurons in the input (first) layer. The architectural graph of Figure 2.3.5 illustrates the layout of a multilayer feedforward neural network for the case of a single hidden layer. For brevity, the network of Figure 2.3.5 is referred to as a 4-3-2 network since it has 4 input neurons, 3 hidden neurons, and 2 output neurons. As another example, a feedforward network with p neurons in the input layer (source nodes), h_1 neurons in the first layer, h_2 neurons in the second layer, and q neurons in the output layer, is referred to as a $p - h_1 - h_2 - q$ network.

The neural network of Figure 2.3.5 is said to be *fully connected* in the sense that

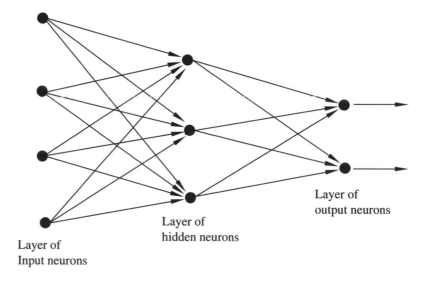

Figure 2.3.5. A fully connected feedforward network with one hidden layer.

every neuron in each layer of the network is connected to every other neuron in the adjacent forward layer. If, however, some of the synaptic connections are missing from the network, we say that the network is *partially connected*. A form of partially connected multilayer feedforward network of particular interest is a *locally connected network*. An example of such a network with a single hidden layer is presented in Figure 2.3.6, where each neuron in the hidden layer is connected to a local (partial) set of input neurons that lies in its immediate neighborhood (such a set of localized neurons feeding a neuron is said to constitute the *receptive field* of the neuron). Likewise, each neuron in the output layer is connected to a local set of hidden neurons.

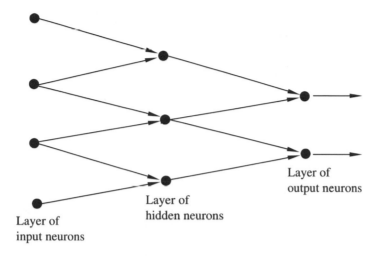

Figure 2.3.6. A partially connected feedforward network.

Another type of network contains feedback, and thus is called a *feedback network* or a *recurrent network*. For example, a recurrent network may consist of a single layer of neurons with each neuron feeding its output signal back to the inputs of all the other neurons, as illustrated in Figure 2.3.7.

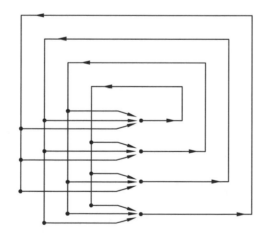

Figure 2.3.7. A recurrent network with no self-feedback loops and no hidden neurons.

In the structure depicted in the figure there is no self-feedback in the network (self-feedback refers to a situation where the output of a neuron is fed back to its own input). The recurrent network illustrated in Figure 2.3.7 also has no hidden neurons. In Figure 2.3.8 we illustrate another class of recurrent networks with hidden neurons.

The feedback connections shown in Figure 2.3.8 originate from the hidden neurons as well as the output neurons. The presence of feedback loops, be it as in the recurrent structure of Figure 2.3.7 or that of Figure 2.3.8, has a profound impact on the learning capability of the network, and on its performance.

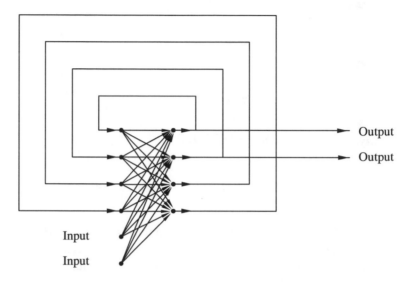

Figure 2.3.8. A recurrent network with hidden neurons.

To conclude this chapter, we note that the performance of a network is determined by many factors including

(1) the internal decay rates;

(2) the external inputs;

(3) the synaptic weights;

(4) the propagation delays;

(5) the signal functions and the thresholds/biases;

(6) the connection topology.

In the deterministic models we discussed, a network with a given connection topology is a dynamical system in the sense that given initial activation levels and initial synaptic coupling coefficients and when all other factors are considered as parameters, the future activation levels and synaptic coupling coefficients can be calculated. This is called the *joint activation-weight dynamics*. In applications, however, one often separates the activation dynamics from the weight dynamics. There are many schemes for adaptively determining the synaptic weights of a network in order to achieve some particular kind of tasks such as pattern classification or to obtain some desired network outputs from a give set of inputs. Such a scheme usually determines a discrete dynamical system (a system of difference equations) or a continuous dynamical system (a system of differential equations) in the space of matrices of synaptic coupling coefficients. This is the *weight dynamics*. Once the connection topology and the synaptic weights are determined and when other factors are regarded as parameters, the future activation levels can be predicted by specifying the initial activation levels. This is the *activation dynamics*. How various factors affect this activation dynamics will be the central subject of the remaining part of this book.

Chapter 3

Simple networks

In this chapter, we present several simple networks that perform some elementary functions such as storing, recalling and recognizing neuron activation patterns. We will discuss a special connection topology which seems to be quite effective in solving the noise-saturation dilemma. Assuming this connection topology, we will also indicate how the choice of signal functions affects the limiting patterns of a network and how to determine the synaptic weights.

3.1   Outstars: pattern learning

An *outstar* is a neural network which is designed for learning and recalling external inputs impressed on an array of neurons. The network consists of $(n+1)$ neurons: the *command neuron* v_0 and the *input neurons* $(v_1, \ldots, v_n)$. The command neuron axon connects with input neurons. Axons of the input neurons are not considered because they do not affect outstar's functioning.

The name "outstar" comes from the geometry when arranging the input neurons in a circle with the command neuron in the center. Figure 3.1.1 gives a schematic

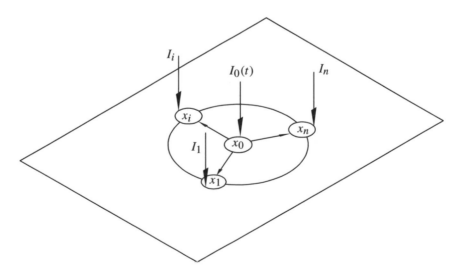

Figure 3.1.1. An outstar neural network.

picture of an outstar. The inputs $(I_1, \ldots, I_n)$ activate the input neurons, and the input I_0 turns on the command neuron, producing axon signals to the input neurons. The external inputs and axon signals modify the LTM traces, storing the input *spatial pattern* (defined later) in the LTM trace. After removing the inputs, the spatial pattern can be restored and recalled across the input neurons by turning on v_0.

Following Harvey [1994], we can formulate a simple mathematical model of the outstar, by using additive STM and passive decay LTM, as follows:

$$\begin{cases} \dot{x}_0(t) = -\alpha x_0(t) + I_0(t), \\ \dot{x}_i(t) = -\alpha x_i(t) + S_i(t)Z_i(t) + I_i(t), \\ \dot{Z}_i(t) = -\beta Z_i(t) + u_i(t)[x_i(t)]^+, \qquad 1 \leq i \leq n, \end{cases} \tag{3.1.1}$$

where α and β are positive constants,

$$\begin{cases} S_i(t) = f(x_0(t-\tau) - \Gamma), \\ u_i(t) = g(x_0(t-\tau) - \Gamma), \end{cases} \quad 1 \leq i \leq n \tag{3.1.2}$$

and $f, g : \mathbb{R} \to \mathbb{R}$ are signal functions with certain positive constant coefficients. Here, for the sake of simplicity, we assumed that this is a "regular" outstar and so the synaptic coupling from the command neuron to the input neurons is uniform in strength, resulting the same signal functions, same transmission delay and the same threshold for each input neuron. For the sake of simplicity, we will consider the case where f and g are smooth functions, and $(I_0, I_1, \ldots, I_n)$ are nonnegative constants. More general situations are discussed in the appendix to this chapter.

The first equation can be easily solved by the variation-of-constants formula

$$x_0(t) = e^{-\alpha t}x_0(0) + \frac{I_0}{\alpha}(1 - e^{-\alpha t}), \quad t \geq 0 \tag{3.1.3}$$

from which it follows that

$$x_0(t) \to \frac{I_0}{\alpha} \quad \text{as } t \to \infty. \tag{3.1.4}$$

Assume $\frac{I_0}{\alpha} > \Gamma$. Substituting this into equations for x_i and Z_i, we then obtain the following linear nonhomogeneous system

$$\begin{cases} \dot{x}_i(t) = -\alpha x_i(t) + f^* Z_i(t) + I_i(t), \\ \dot{Z}_i(t) = -\beta Z_i(t) + g^* x_i(t), \end{cases} \quad 1 \leq i \leq n \tag{3.1.5}$$

as the system to describe the limiting behaviors of nonnegative solutions $(x_1(t), \ldots, x_n(t))$ and $(Z_1(t), \ldots, Z_n(t))$ as $t \to \infty$, where

$$\begin{cases} f^* = f\left(\frac{I_0}{\alpha} - \Gamma\right), \\ g^* = g\left(\frac{I_0}{\alpha} - \Gamma\right). \end{cases} \tag{3.1.6}$$

It is interesting to note that (3.1.5) is a decoupled n systems of planar equations, and it is easy to show that if

$$\alpha\beta > f^*g^* \tag{3.1.7}$$

then

$$x_i(t) \to x_i^* = \frac{\beta}{\alpha\beta - f^*g^*} I_i, \quad Z_i(t) \to Z_i^* = \frac{g^*}{\alpha\beta - f^*g^*} I_i \quad \text{as } t \to \infty. \tag{3.1.8}$$

We also note that for the *total activity*, *total LTM trace* and *total input* given respectively by

$$X(t) = \sum_{i=1}^n x_i(t), \quad Z(t) = \sum_{i=1}^n Z_i(t), \quad I(t) = \sum_{i=1}^n I_i(t), \tag{3.1.9}$$

we have

$$\begin{cases} \dot{X}(t) = -\alpha X(t) + f^*Z(t) + I(t), \\ \dot{Z}(t) = -\beta Z(t) + g^*X(t), \end{cases} \tag{3.1.10}$$

from which it follows that

$$X(t) \to \frac{\beta}{\alpha\beta - f^*g^*} I, \quad Z(t) \to \frac{g^*}{\alpha\beta - f^*g^*} I \quad \text{as } t \to \infty. \tag{3.1.11}$$

For further discussions, it is convenient to introduce the so-called *reflectance coefficients* $(\theta_1, \ldots, \theta_n)$. These are nonnegative constants such that

$$I_i(t) = \theta_i I(t) = \theta_i \sum_{j=1}^n I_j. \tag{3.1.12}$$

We say that these coefficients define a *spatial pattern*. Such a spatial pattern is important in describing neural activities. For example, in vision the identification of a picture is invariant under fluctuations in total intensity $I(t)$ over a broad physiological range, and the relative intensity θ_i at each spatial point characterizes the picture. Convergence (3.1.8) shows that the steady-state STM trace and LTM trace are both proportional to the input spatial pattern. Equation (3.1.10) and convergence (3.1.11), however, show that the total STM trace $X(t)$ and the total LTM trace $Z(t)$ are completely determined by the total intensity of the external inputs to input neurons.

In summary, in outstar *learning*, one applies I_0 and $(I_1, \ldots, I_n)$ to the network: I_0 excites v_0 causing a (sampling) signal to the input neurons. The impact of this sample signal and the applied spatial pattern cause the synaptic coupling relax towards its steady state, storing the pattern. When the sampling signal and external inputs are removed, the LTM trace retains the pattern.

To *recall* the stored pattern, one activates the command neuron v_0 with a new input $\tilde{I}_0$. Without external inputs to input neurons, the STM for v_i becomes

$$\dot{x}_i(t) = -\alpha x_i(t) + f(x_0(t - \tau) - \Gamma)Z_i^*$$

and thus ν_i relaxes to

$$x_i(t) \to \frac{1}{\alpha} f\left(\frac{\tilde{I}_0}{\alpha} - \Gamma\right) Z_i^* = \frac{g^* f\left(\frac{\tilde{I}_0}{\alpha} - \Gamma\right)}{\alpha(\alpha\beta - f^*g^*)} \theta_i I \quad \text{as } t \to \infty.$$

In particular, $\lim_{t\to\infty} \frac{x_i(t)}{x_j(t)} = \frac{\theta_i}{\theta_j}$ for all $i, j = 1, \ldots, n$. That is, the readout signal recalls the spatial pattern on the input neurons, regardless of the activating new input to the command neuron and regardless of the initial state of the input neurons.

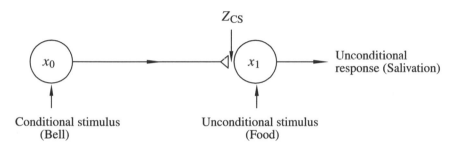

Figure 3.1.2. A simple outstar simulating Pavlovian conditioning.

The above discussion gives insights into the learning mechanism of *Pavlovian conditioning*. A famous example of Pavlovian conditioning is as follows: A hungry dog, presented with food (the *unconditional stimulus*, or UCS) will salivate (the *unconditional response*, or UCR). A bell (the *conditional stimulus*, or CS) does not initially elicit salivation, but will do after pairing CS and UCS several times. To apply the above discussion, assume a neuron ν_0 is activated by CS and a neuron ν_1 is activated by UCS. Then the synaptic coupling Z_{CS} between ν_0 and ν_1 increases and eventually a CS alone produces the UCR.

The remaining part of this section provides a rigorous proof, taken from Huang and Wu [1999a], of the convergence result (3.1.8) for system (3.1.1). For more general results, called Grossberg's Learning Theorem, we refer to the appendix.

We first recall the *Gronwall's inequality* stated as follows.

Lemma 3.1.1. *Suppose that $t_0 \in \mathbb{R}$ and that $\varphi, \psi, w, \xi : [t_0, \infty) \to \mathbb{R}$ are continuous functions with $w(t), \xi(t) \geq 0$ for all $t \geq t_0$. If*

$$\varphi(t) \leq \psi(t) + \xi(t) \int_{t_0}^{t} w(s)\varphi(s)\,\mathrm{d}s \quad \text{for } t \geq t_0, \tag{3.1.13}$$

then

$$\varphi(t) \leq \psi(t) + \xi(t) \int_{t_0}^{t} \psi(s)w(s)e^{\int_s^t \xi(u)w(u)\,\mathrm{d}u}\,\mathrm{d}s \quad \text{for } t \geq t_0. \tag{3.1.14}$$

To prove Lemma 3.1.1, we multiply (3.1.13) by $w(t)$ and obtain for $t \geq t_0$,

$$\frac{d}{dt} \left[e^{-\int_{t_0}^{t} \xi(s)w(s)\,ds} \int_{t_0}^{t} \varphi(s)w(s)\,ds \right] \leq \psi(t)w(t)e^{-\int_{t_0}^{t} \xi(s)w(s)\,ds}.$$

Thus integration yields

$$e^{-\int_{t_0}^{t} \xi(s)w(s)\,ds} \int_{t_0}^{t} \varphi(s)w(s)\,ds \leq \int_{t_0}^{t} \psi(s)w(s)e^{-\int_{t_0}^{s} \xi(u)w(u)\,du}\,ds,$$

or equivalently

$$\int_{t_0}^{t} \varphi(s)w(s)\,ds \leq \int_{t_0}^{t} \psi(s)w(s)e^{\int_{s}^{t} \xi(u)w(u)\,du}\,ds.$$

Substituting this into (3.1.13), we obtain (3.1.14).

Lemma 3.1.2. *Let $t_0 \in \mathbb{R}$, P, Q, F, G, I : $[t_0, \infty) \to \mathbb{R}$ be continuous with I being bounded and*

$$P(t),\, Q(t),\, F(t),\, G(t) \to 0 \quad as\ t \to \infty.$$

Assume that α, β, δ and γ are nonnegative constants such that the real parts of all eigenvalues of $A = \begin{pmatrix} -\alpha & \delta \\ \gamma & -\beta \end{pmatrix}$ are negative. Then every solution of

$$\begin{cases} \dot{x} = -\alpha x + \delta z + P(t)x + F(t)z + I(t), \\ \dot{z} = \gamma x - \beta z + G(t)x + Q(t)z \end{cases} \tag{3.1.15}$$

is bounded on $[t_0, \infty)$.

Proof. Let $y(t) = \begin{pmatrix} x(t) \\ z(t) \end{pmatrix}$ be a given solution of (3.1.15) and let $X(t) = \begin{pmatrix} P(t) & F(t) \\ G(t) & Q(t) \end{pmatrix}$ for $t \geq t_0$. Denote by $\| \cdot \|$ the Euclidean norm in $\mathbb{R}^2$, and for a 2×2 real matrix B let $\|B\|$ be the matrix norm so that $\left\| B \begin{pmatrix} x \\ z \end{pmatrix} \right\| \leq \|B\| \left\| \begin{pmatrix} x \\ z \end{pmatrix} \right\|$ for all $\begin{pmatrix} x \\ z \end{pmatrix} \in \mathbb{R}^2$. Applying the variation-of-constants formula to (3.1.15), we get

$$\|y(t)\| \leq \|e^{A(t-t_0)}\| \left\| \begin{pmatrix} x(t_0) \\ z(t_0) \end{pmatrix} \right\| + \left\| \int_{t_0}^{t} e^{A(t-s)} \begin{pmatrix} I(s) \\ 0 \end{pmatrix} ds \right\|$$

$$+ \left\| \int_{t_0}^{t} e^{A(t-s)} X(s)y(s)\,ds \right\|, \quad t \geq t_0.$$

We can find constants K, $\varepsilon > 0$ so that

$$\|e^{At}\| \leq Ke^{-\varepsilon t} \quad \text{for } t \geq 0.$$

Therefore,

$$\|y(t)\| \le K e^{-\varepsilon(t-t_0)} \left\| \begin{pmatrix} x(t_0) \\ z(t_0) \end{pmatrix} \right\| + K \int_{t_0}^{t} e^{-\varepsilon(t-s)} |I(s)| \, ds$$

$$+ K \int_{t_0}^{t} e^{-\varepsilon(t-s)} \|X(s)\| \|y(s)\| \, ds, \quad t \ge t_0.$$

Let

$$C = \sup_{t \ge t_0} \left[K e^{-\varepsilon(t-t_0)} \left\| \begin{pmatrix} x(t_0) \\ y(t_0) \end{pmatrix} \right\| + K \int_{t_0}^{t} e^{-\varepsilon(t-s)} |I(s)| \, ds \right].$$

Clearly, $C < \infty$ and Lemma 3.1.1 implies

$$\|y(t)\| \le C + KC \int_{t_0}^{t} e^{-\varepsilon(t-s)} \|X(s)\| e^{K \int_s^t \|X(u)\| \, du} \, ds$$

$$= C + KC \int_{t_0}^{t} e^{\int_s^t [-\varepsilon + K\|X(u)\|] \, du} \|X(s)\| \, ds.$$

Since $\|X(t)\| \to 0$ as $t \to \infty$, we can find $T \ge t_0$ so that

$$\|X(u)\| < 1 \quad \text{and} \quad -\varepsilon + K\|X(u)\| < -\frac{\varepsilon}{2} \quad \text{for } u \ge T.$$

Therefore, for $t \ge T$, we have

$$\|y(t)\| \le C + KC \left[\int_{t_0}^{T} e^{\int_s^t [-\varepsilon + K\|X(u)\|] \, du} \|X(s)\| \, ds \right.$$

$$\left. + \int_{T}^{t} e^{\int_s^t [-\varepsilon + K\|X(u)\|] \, du} \|X(s)\| \, ds \right]$$

$$\le C + KC e^{\int_T^t [-\varepsilon + K\|X(u)\|] \, du} \int_{t_0}^{T} e^{\int_s^T [-\varepsilon + K\|X(u)\|] \, du} \|X(s)\| \, ds$$

$$+ KC \int_{T}^{t} e^{-\frac{\varepsilon}{2}(t-s)} \, ds$$

$$\le C + KC \int_{t_0}^{T} e^{\int_s^T [-\varepsilon + K\|X(u)\|] \, du} \|X(s)\| \, ds + \frac{2KC}{\varepsilon}.$$

This proves the boundedness of y on $[t_0, \infty)$.

Lemma 3.1.3. *Assume that all the all conditions of Lemma 3.1.2 are satisfied. Let* $\begin{pmatrix} X \\ Z \end{pmatrix} : [t_0, \infty) \to \mathbb{R}^2$ *be the solution of*

$$\begin{cases} \dot{X} = -\alpha X + \delta Z + I(t), \\ \dot{Z} = \gamma X - \beta Z \end{cases} \tag{3.1.16}$$

with $X(t_0) = x(t_0)$ and $Z(t_0) = z(t_0)$. Then

$$\lim_{t \to \infty} [X(t) - x(t)] = \lim_{t \to \infty} [Z(t) - z(t)] = 0.$$

Proof. Using the variation-of-constants formula to (3.1.15) and (3.1.16) and the fact that $X(t_0) - x(t_0) = Z(t_0) - z(t_0) = 0$, we obtain

$$\begin{pmatrix} X(t) - x(t) \\ Z(t) - z(t) \end{pmatrix} = - \int_{t_0}^{t} e^{A(t-s)} \begin{bmatrix} P(s) & F(s) \\ G(s) & Q(s) \end{bmatrix} \begin{pmatrix} x(s) \\ z(s) \end{pmatrix} ds.$$

By Lemma 3.1.2, we can find a constant $K_1 > 0$ such that

$$\left\| \begin{pmatrix} X(t) - x(t) \\ Z(t) - z(t) \end{pmatrix} \right\| \le K K_1 \int_{t_0}^{t} e^{-\varepsilon(t-s)} \left\| \begin{bmatrix} P(s) & F(s) \\ G(s) & Q(s) \end{bmatrix} \right\| ds$$

$$\to 0 \quad \text{as } t \to \infty$$

since $\left\| \begin{pmatrix} P(s) & F(s) \\ G(s) & Q(s) \end{pmatrix} \right\| \to 0$ as $s \to \infty$. This completes the proof.

To apply the above results to (3.1.1), we note that $x_0(t) \to \frac{I_0}{\alpha}$ as $t \to \infty$, and hence

$$S_i(t) \to f^*, \quad u_i(t) \to g^* \qquad \text{as } t \to \infty.$$

We can then write (3.1.1) as (3.1.15) with

$$\begin{array}{llll} x(t) = x_i(t), & z(t) = Z_i(t), & \delta = f^*, & \gamma = g^*, \\ F(t) = S_i(t) - f^*, & P(t) = 0, & G(t) = u_i(t) - g^*, & Q(t) = 0. \end{array}$$

(3.1.5) is exactly the associated "limiting" equation (3.1.16) of (3.1.15), and condition (3.1.7) and Lemma 3.1.3 then ensure that

$$\lim_{t \to \infty} x_i(t) = x_i^*, \quad \lim_{t \to \infty} z_i(t) = Z_i^* \qquad \text{for } 1 \le i \le n.$$

3.2   Instars: pattern recognition

An *instar* is a neural network for recognizing spatial patterns. As with outstars, the name comes from the geometry. The network consists of n input neurons ($v_1, \ldots, v_n$) and one output neuron v_0. When inputs ($I_1, \ldots, I_n$) are imposed, input neurons are activated and send signals to the output neuron. The model for the network shown in Figure 3.2.1 during the *learning* process takes the following form

$$\begin{cases} \dot{x}_i(t) = -\alpha x_i(t) + I_i(t), & 1 \le i \le n, \\ \dot{x}_0(t) = -\alpha x_0(t) + \sum_{i=1}^{n} f(x_i(t - \tau_i) - \Gamma_i)Z_i(t) + I_0(t), & (3.2.1) \\ \dot{Z}_i(t) = -\beta Z_i(t) + g(x_i(t - \tau_i) - \Gamma_i)[x_0]^+, \end{cases}$$

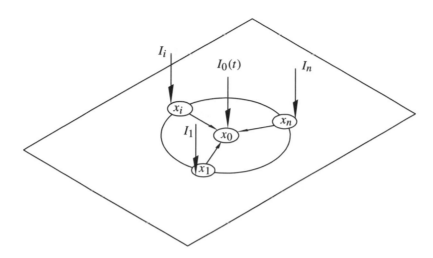

Figure 3.2.1. An instar during a learning process.

where $f, g : \mathbb{R} \to [0, \infty)$ are given signal functions.

As for outstars, we consider the simplest case where $(I_1, \ldots, I_n)$ and I_0 are constant functions and define

$$\theta_i = \frac{I_i}{I}, \quad 1 \leq i \leq n \tag{3.2.2}$$

as the *spatial pattern* (reflectance coefficients), where $I = \sum_{i=1}^{n} I_i$. The system which determines the limiting behaviors of $(x_0, Z_1, \ldots, Z_n)$ as $t \to \infty$ is, assuming nonnegative x_0 and large inputs such that

$$\frac{I_i}{\alpha} > \Gamma_i, \quad 1 \leq i \leq n \tag{3.2.3}$$

of the form

$$\begin{cases} \dot{x}_0 = -\alpha x_0 + \sum_{i=1}^{n} f\left(\frac{I_i}{\alpha} - \Gamma_i\right) Z_i + I_0, \\ \dot{Z}_i = -\beta Z_i + g\left(\frac{I_i}{\alpha} - \Gamma_i\right) x_0. \end{cases} \tag{3.2.4}$$

In particular, for the *weighted sum* of LTM trace

$$Z(t) = \sum_{i=1}^{n} f\left(\frac{I_i}{\alpha} - \Gamma_i\right) Z_i$$

we have

$$\begin{cases} \dot{x}_0 = -\alpha x_0 + Z(t) + I_0, \\ \dot{Z} = -\beta Z + \left[\sum_{i=1}^{n} f\left(\frac{I_i}{\alpha} - \Gamma_i\right) g\left(\frac{I_i}{\alpha} - \Gamma_i\right)\right] x_0, \end{cases} \tag{3.2.5}$$

from which it follows that, if

$$\alpha\beta > \sum_{i=1}^{n} f\left(\frac{I_i}{\alpha} - \Gamma_i\right) g\left(\frac{I_i}{\alpha} - \Gamma_i\right) \tag{3.2.6}$$

then

$$x_0(t) \rightarrow \frac{\beta I_0}{\alpha\beta - \sum\limits_{i=1}^{n} f\left(\frac{I_i}{\alpha} - \Gamma_i\right) g\left(\frac{I_i}{\alpha} - \Gamma_i\right)},$$

$$Z(t) \rightarrow \frac{I_0 \sum\limits_{i=1}^{n} f\left(\frac{I_i}{\alpha} - \Gamma_i\right) g\left(\frac{I_i}{\alpha} - \Gamma_i\right)}{\alpha\beta - \sum\limits_{i=1}^{n} f\left(\frac{I_i}{\alpha} - \Gamma_i\right) g\left(\frac{I_i}{\alpha} - \Gamma_i\right)} \qquad \text{as } t \rightarrow \infty. \tag{3.2.7}$$

Substituting this into (3.2.4), we then obtain

$$Z_i(t) \rightarrow \frac{g\left(\frac{I_i}{\alpha} - \Gamma_i\right)}{\alpha\beta - \sum\limits_{i=1}^{n} f\left(\frac{I_i}{\alpha} - \Gamma_i\right) g\left(\frac{I_i}{\alpha} - \Gamma_i\right)} I_0 = Z_i^* \qquad \text{as } t \rightarrow \infty. \tag{3.2.8}$$

In particular,

$$\frac{Z_i(t)}{Z_j(t)} \rightarrow \frac{g\left(\frac{I_i}{\alpha} - \Gamma_i\right)}{g\left(\frac{I_j}{\alpha} - \Gamma_j\right)} \qquad \text{as } t \rightarrow \infty. \tag{3.2.9}$$

If we define the *modified reflectance coefficients* as

$$\theta_i^* = \frac{g\left(\frac{I_i}{\alpha} - \Gamma_i\right)}{\sum\limits_{j=1}^{n} g\left(\frac{I_j}{\alpha} - \Gamma_j\right)}, \qquad 1 \le i \le n, \tag{3.2.10}$$

then the steady-state LTM vector is proportional to the modified input pattern expressed as the modified reflectance coefficient vector $(\theta_1^*, \ldots, \theta_n^*)$. That is, the instar learns the pattern and stores it in the LTM trace, as in outstars. Note that θ^* coincides with the reflectance coefficient vector $\theta = (\theta_1, \ldots, \theta_n)$ if $\Gamma_i = 0$ for $1 \le i \le n$ and if g is a linear function in the neighborhood of $\frac{I_i}{\alpha}$ for $1 \le i \le n$.

An important function of an instar is *pattern recognition*. Instar recognition is by comparing an input pattern to the stored LTM vector. To be specific, we assume a new input pattern

$$P = (p_1, \ldots, p_n) \tag{3.2.11}$$

is imposed on the input neuron vector. Then the limiting behavior of the STM of the output neuron with the aforementioned stored pattern $(Z_1^*, \ldots, Z_n^*)$ is governed by

$$\dot{x}_0 = -\alpha x_0 + \sum\limits_{i=1}^{n} f\left(\frac{p_i}{\alpha} - \Gamma_i\right) Z_i^* \tag{3.2.12}$$

and thus the output is given by

$$x_0^* = \frac{1}{\alpha} \sum\limits_{i=1}^{n} f\left(\frac{p_i}{\alpha} - \Gamma_i\right) Z_i^* = \frac{I_0}{\alpha} \sum\limits_{i=1}^{n} \frac{f\left(\frac{p_i}{\alpha} - \Gamma_i\right) g\left(\frac{I_i}{\alpha} - \Gamma_i\right)}{\alpha\beta - \sum\limits_{i=1}^{n} f\left(\frac{p_i}{\alpha} - \Gamma_i\right) g\left(\frac{I_i}{\alpha} - \Gamma_i\right)}. \tag{3.2.13}$$

Assume this output is large enough to excess the threshold, namely,

$$x_0^* > \Gamma_0, \tag{3.2.14}$$

then the neuron fires, thus *recognizing the pattern.* We thus say that *P belongs to the pattern class represented by* $(Z_1^*, \ldots, Z_n^*)$ *if* (3.2.14) *is satisfied.* Note that if all Γ_i are zero and if both f and g are linear, this is equivalent to requiring the inner product of the input pattern P and the stored pattern θ

$$\theta \cdot P = \sum_{i=1}^{n} \theta_i \, p_i \tag{3.2.15}$$

to be sufficiently large.

Figure 3.2.2 gives a schematic description of instar pattern recognition.

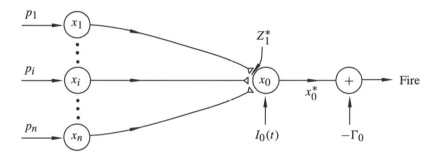

Figure 3.2.2. Recognition of a pattern by an instar with stored LTM trace $(Z_1^*, \ldots, Z_n^*)$: the inputs cause input neurons to send signals along the axons to the output neuron. The signals are multiplied by the LTM trace, summed and compared with a threshold. If the output neuron fires, the input pattern is recognized as belonging to the pattern class represented by the LTM trace.

In summary, instars and outstars are dual to one another. An outstar can recall but cannot recognize a pattern, while an instar can recognize but cannot recall. In other words, the outstar is blind and the instar is dumb.

Admittedly, the models for outstars and instars we discussed in the last two sections represent oversimplification of the reality for even idealized networks. Interested readers are referred to the appendix where general results of Grossberg for unbiased pattern learning are described. See also Carpenter [1994] for more complicated spatial pattern learning by a distributed outstar where the source node (command neuron) can be a source field of arbitrarily many nodes whose activity pattern may be arbitrarily distributed or compressed.

3.3   Lateral inhibition: noise-saturation dilemma

One of the main motivations behind the development of *shunting networks* is to prevent too high or too low activity of either a single neuron or the whole network. This is related to the so-called *noise-saturation dilemma*: Suppose that STM traces at a network level fluctuate within fixed finite limits at their respective network nodes. If a large number of intermitted input sources converge to the nodes through time, then a serious design problem arises due to the fact that the total input converging to each node can vary widely through time. If the STM traces are sensitive to large inputs, then why do not small inputs get lost in internal system noise? If the STM traces are sensitive to small inputs, then why do they not all saturate at their maximum values in response to large inputs?

Shunting networks possess automatic gain control properties which are capable of generating an infinite dynamic range within which input patterns can be effectively processed, thereby solving the noise-saturation dilemma. In this and next sections, we are going to follow Grossberg [1988, 1982] to describe the simplest feedforward and feedback networks to convey some of the main ideas behind shunting networks.

The main focus of this section is on the simplest feedforward network in order to illustrate how it offers a solution to the sensitivity problem raised by the noise-saturation dilemma. Let a spatial pattern $I_i = \theta_i I$ of inputs be processed by the neurons v_i, $i = 1, \ldots, n$. Each θ_i is the (constant) reflectance of its input I_i and I is the (variable) total input size. How can each neuron v_i maintain its sensitivity to θ_i when I is increased? How is saturation avoided?

Note that to compute $\theta_i = \frac{I_i}{\sum_{k=1}^{n} I_k}$, each neuron v_i must have information about all the inputs I_k, $k = 1, \ldots, n$. Rewriting

$$\theta_i = \frac{I_i}{I_i + \sum_{k \neq i} I_k},$$

we observe that increasing I_i increases θ_i whereas increasing any I_k with $k \neq i$ decreases θ_i. Translating this into the connection topology for delivering feedforward inputs to the neuron v_i suggests that I_i excites v_i and that all I_k with $k \neq i$ inhibit v_i. This rule represents the simplest feedforward *on-center off-surround network* shown in Figure 3.3.1.

Experimental studies of the hippocampus (Andersen, Gross, Lomo and Sveen [1969]) have suggested that its cells are arranged in a recurrent on-center off-surround anatomy. The main cell type, the pyramidal cell, emits axon collaterals to interneurons. Some of these interneurons feed back excitatory signals to nearby pyramidal cells. Other interneurons scatter inhibitory feedback signals over a broad area.

How does the on-center off-surround network activate and inhibit the neuron v_i via mass action? Assume for each v_i the STM trace or activation at time t is $x_i(t)$ whose maximum is B. As discussed in Chapter 2, one may think of B as the maximal excitable site among which $x_i(t)$ are excited and $B - x_i(t)$ are unexcited at the time t. Then at v_i, I_i excites $B - x_i$ unexcited sites by mass action, and the total inhibitory

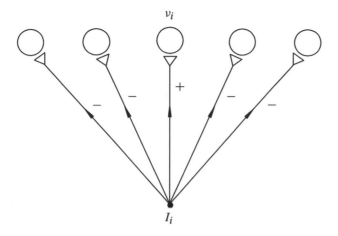

Figure 3.3.1. An on-center off-surround network.

input $\sum_{k \neq i} I_k$ inhibits x_i excited sites by mass action. Assume also the excitation decays at a fixed rate A. Then one gets

$$\frac{d}{dt}x_i = -Ax_i + (B - x_i)I_i - x_i \sum_{k \neq i} I_k, \quad 1 \leq i \leq n. \qquad (3.3.1)$$

If a fixed spatial pattern $I_i = \theta_i I$, $1 \leq i \leq n$, is presented and the background input is held constant for a while, from the equivalent form

$$\frac{d}{dt}x_i = -(A + I)x_i + BI_i, \quad 1 \leq i \leq n \qquad (3.3.2)$$

of (3.3.1), we conclude that each x_i approaches an equilibrium given by

$$x_i^* = \theta_i \frac{BI}{A + I}, \quad 1 \leq i \leq n. \qquad (3.3.3)$$

Consequently, the *relative activity* $X_i^* = x_i^* / \sum_{k=1}^{n} x_k^*$ equals the reflectance θ_i no matter how large I is chosen and the *total activity*

$$X^* = \sum_{i=1}^{n} x_i^* = \frac{BI}{A + I} \qquad (3.3.4)$$

is bounded by B, so there is no saturation. This regulation of the network's total activation, called *total activity normalization*, is due to the automatic control by the inhibitory inputs $(-x_i \sum_{k \neq i} I_k)$.

The steady state in (3.3.3) combines two types of information: information about spatial pattern θ_i and information about background activity or *luminance I*. Note that formula (3.3.4) shows that the total activity is independent of the number of active neurons. This normalization rule is a conservation law which says, for example, that in a network that receives a fixed total luminance, making one part of the field brighter tends to make another part of the field darker. This property helps to explain *brightness contrast*.

We conclude this section by recalling that equation (3.3.1) is a special case of the membrane equation describing the voltage $V(t)$ of a neuron

$$C\frac{dV}{dt} = (V^+ - V)g^- + (V^- - V)g^+ + (V^P - V)g^P, \qquad (3.3.5)$$

where, as described in Section 2.1, C is the capacitance, V^+, V^- and V^P are constant excitatory, inhibitory and passive saturation points, respectively, and g^+, g^- and g^P are excitatory, inhibitory and passive conductances, respectively. To obtain (3.3.1) from (3.3.5), we suppose (3.3.5) holds at each neuron v_i and let, at v_i, $V = x_i$, $C = 1$ (rescale time), $V^+ = B$, $V^- = V^P = 0$, $g^+ = I_i$, $g^- = \sum_{k \neq i} I_k$ and $g^P = A$.

3.4   Recurrent ON-CTR OFF-SUR networks: signal enhancement and noise suppression

Recurrent on-center off-surround networks have been found in a variety of neural structures such as neocortex (Stefanis [1969]) and cerebellum (Eccles, Ito and Szentagothai [1967]) and hippocampus (Andersen, Gross, Lomo and Sveen [1969]).

Consider a network of population of neurons $v_1, \ldots, v_n$. Let the average excitation at time t of the ith population v_i be denoted by $x_i(t)$. We want to study how these averages are transformed through time by recurrent on-center off-surround interactions via the system of equations

$$\dot{x}_i = -Ax_i + (B - x_i)f(x_i) - x_i \sum_{k \neq i} f(x_k) + I_i, \quad 1 \leq i \leq n.$$

Four effects which determine this system are as follows:

(i) exponential decay via the term $-Ax_i$;

(ii) shunting self-excitation via the term $(B - x_i)f(x_i)$, where B is the maximal excitation level;

(iii) shunting inhibition of other populations via the term $-x_i \sum_{k \neq i} f(x_k)$;

(iv) externally applied inputs via the term I_i.

See Figure 3.4.1 for a schematic picture.

The nonnegative function $f(w)$ with $f(0) = 0$ describes the mean output signal of a given population as a function of its activity w. This function is determined by such

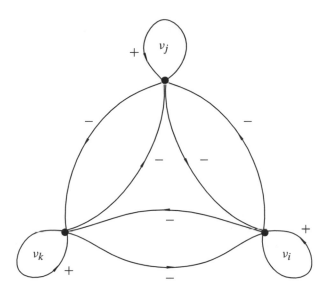

Figure 3.4.1. A recurrent on-center off-surround network.

parameters as the distribution of signal (or spiking) thresholds and of afferent synapses per neuron within each population (Wilson and Cowan [1972]), some examples were given in Section 2.2, but other forms are possible. We are going to see that varying the function f can dramatically changes the pattern features stored by the network in STM.

   We will always set the external inputs I_i, $1 \le i \le n$, equal to zero, yielding the nonlinear system

$$\dot{x}_i = -\left[A + \sum_{k \ne i} f(x_k)\right] x_i + (B - x_i) f(x_i), \quad 1 \le i \le n. \tag{3.4.1}$$

In practice, $I = (I_1, \ldots, I_n)$ is switched on during a finite time interval $[-\tau, 0]$. Given prescribed initial data at $t = -\tau$, these inputs will determine a particular distribution of terminal values $x_i(0)$, $1 \le i \le n$. These will serve as the initial values of system (3.4.1). Our goal is to determine how these initial values are transformed as $t \to \infty$.

   The main focus of this section is to show how the choice of f affects the answers to the following two questions:

(a) Under what circumstances, is the reverberation persistent? transient?

(b) How is the initial pattern of activity, that was laid down by previous external inputs, transformed as time goes on? Is there a limiting pattern of activity, or does the pattern oscillate indefinitely?

To be more specific, we first introduce two important variables: the *total activity*

$$x(t) = \sum_{i=1}^{n} x_i(t) \tag{3.4.2}$$

and the *pattern variable*

$$X_i(t) = x_i(t)/x(t), \quad 1 \le i \le n. \tag{3.4.3}$$

Throughout the remaining discussion, we are going to assume that $f : \mathbb{R} \to \mathbb{R}$ is locally Lipschitz continuous. So the Cauchy initial value problem (3.4.1) has a unique solution. See, for example, Hale [1969].

Note that if $x_i(t_0) = 0$ for some $t_0 \in \mathbb{R}$, then $x_i(t) = 0$ for all t where the solution $(x_1(t), \ldots, x_n(t))$ of (3.4.1) is defined. In particular, $x_i(0) = 0$ implies $x_i(t) = 0$ for all $t \ge 0$ where $(x_1(t), \ldots x_n(t))$ is defined, so such a v_i can be deleted from the network without loss of generality. Therefore, we will always assume

$$B \ge x_i(0) > 0 \tag{3.4.4}$$

for each $1 \le i \le n$. Under this assumption, $x_i(t) > 0$ for $1 \le i \le n$ and $t \ge 0$. Using equation (3.4.1), we can also show that[1]

$$0 < x_i(t) \le B \quad \text{for } 1 \le i \le n, t \ge 0. \tag{3.4.5}$$

Consequently, both the pattern variable and the total activity are well defined for all $t \ge 0$.

We say the reverberation is *transient* if $\lim_{t\to\infty} x(t) = 0$, and *persistent* if $\liminf_{t\to\infty} x(t) > 0$ for every solution of (3.4.1) satisfying the initial condition (3.4.4). We introduce

$$g(w) = \frac{f(w)}{w} \quad \text{for } w > 0 \tag{3.4.6}$$

and

$$G(x_1, \ldots, x_n) = \sum_{k=1}^{n} X_k g(x_k). \tag{3.4.7}$$

Clearly,

$$G(x_1, \ldots, x_k) = x^{-1} \sum_{k=1}^{n} f(x_k). \tag{3.4.8}$$

---

[1]The following is a well-known fact to determine the invariance of the interval $I = [a_1, b_1] \times [a_2, b_2] \times \cdots \times [a_n, b_n]$ in $\mathbb{R}^n$ for the flow defined by a system of ordinary differential equations $\dot{x}_i(t) = h_i(x_1(t), \ldots x_n(t))$ with locally Lipschitz continuous mappings $h_i : \mathbb{R}^n \to \mathbb{R}, 1 \le i \le n$: Assume that $h_i(x_1, \ldots x_n) \ge 0$ for any $(x_1, \ldots, x_n) \in I$ with $x_i = a$, and $h_i(x_1, \ldots, x_n) \le 0$ for any $(x_1, \ldots, x_n) \in I$ with $x_i = b$. Then $(x_1(0), \ldots, x_n(0)) \in I$ implies that $(x_1(t), \ldots x_n(t)) \in I$ for all $t \ge 0$. See Smith [1995].

Note that (3.4.1) can be written as

$$\dot{x}_i = Bf(x_i) - \left[A + \sum_{k=1}^{n} f(x_k)\right] x_i, \quad 1 \le i \le n$$

from which it follows that

$$\dot{x} = B \sum_{k=1}^{n} f(x_k) - \left[A + \sum_{k=1}^{n} f(x_k)\right] x.$$

Consequently, we have the equation for the total activity

$$\dot{x} = -Ax + (B - x)xG. \tag{3.4.9}$$

It is easy to see that there must be $t_0 \ge 0$ so that $x(t_0) < B$, for otherwise, (3.4.9) implies that $\dot{x}(t) \le -Ax(t)$ for all $t \ge 0$ from which it follows that $x(t) \le e^{-At}x(0) \to 0$ as $t \to \infty$. (3.4.9) also implies that if $x(t_0) < B$ then $x(t) < B$ for $t \ge t_0$. This is because, if otherwise, then there exists $t^* > t_0$ so that $\dot{x}(t^*) \ge 0$ and $x(t^*) = B$, but (3.4.9) implies $\dot{x}(t^*) = -Ax(t^*) < 0$. Consequently, in our discussion about limiting behaviors of $x(t)$ and $X_i(t)$, we will always assume that $0 < x(t) < B$ for all $t \ge 0$.

Note also that $X_i = x_i x^{-1}$ implies $\dot{X}_i = x^{-1}[\dot{x}_i - X_i \dot{x}]$, from which it follows that

$$\dot{X}_i = x^{-1}\left\{Bf(x_i) - \left[A + \sum_{k=1}^{n} f(x_k)\right] x_i - X_i B \sum_{k=1}^{n} f(x_k) + x_i\left[A + \sum_{k=1}^{n} f(x_k)\right]\right\}$$

$$= x^{-1}\left[Bf(x_i) - BX_i \sum_{k=1}^{n} f(x_k)\right].$$

Therefore, we have the following equation for the pattern variables

$$\dot{X}_i = BX_i[g(x_i) - G], \quad 1 \le i \le n. \tag{3.4.10}$$

Using the fact that $\sum_{k=1}^{n} X_k = 1$, we have

$$\dot{X}_i = BX_i \sum_{k=1}^{n} X_k[g(x_i) - g(x_k)], \quad 1 \le i \le n \tag{3.4.11}$$

or

$$\dot{X}_i = BX_i \sum_{k=1}^{n} X_k[g(X_i x) - g(X_k x)], \quad 1 \le i \le n. \tag{3.4.12}$$

For a given solution $(x_1, \ldots, x_n)$ of (3.4.1), we define

$$Q_i = \lim_{t \to \infty} X_i(t), \quad 1 \le i \le n,$$
$$E = \lim_{t \to \infty} x(t)$$

if the limits exist.

Definition 3.4.1. The limiting distribution is *fair* if $Q_i = X_i(0)$ for $1 \leq i \leq n$; *uniform* if $Q_i = \frac{1}{n}$ for $1 \leq i \leq n$; *locally uniform* if $Q_{i_j} = \frac{1}{m}$ for $j = 1, \ldots, m$ and for some $1 < m < n$; *0-1 type* if $Q_i = 1$ for some $1 \leq i \leq n$; *trivalent* if each Q_i assumes one of three given values. We say the limiting distribution exhibits *quenching* if $Q_j = 0$ for some $1 \leq j \leq n$.

In the following definitions, we let

$$M(t) = \max\{X_i(t); i = 1, \ldots, n\},$$
$$m(t) = \min\{X_i(t); i = 1, \ldots, n\}.$$

Definition 3.4.2. The limiting distribution *exhibits contour enhancement* if $\dot{M}(t) \geq 0$ and $\dot{m}(t) \leq 0$, and neither of these derivatives is identically zero. The limiting distribution is *uniformized* if $\dot{M}(t) \leq 0$ and $\dot{m}(t) \geq 0$, and neither of these derivatives is identically zero.

Finally, we introduce

Definition 3.4.3. The reverberation is *normalized* if there exists a unique positive E^* such that $\lim_{t \to \infty} x(t) = E^*$ for every solution of (3.4.1) satisfying the initial condition (3.4.4).

The following result demonstrates the "order-preserving" property for the pattern variables.

Proposition 3.4.4. *Suppose that the pattern variables X_i are labeled in such a way that $X_1(0) \leq X_2(0) \leq \cdots \leq X_n(0)$. Then $X_1(t) \leq X_2(t) \leq \cdots \leq X_n(t)$ for all $t \geq 0$.*

Proof. If the conclusion fails, that is, there exist $i \in \{1, \ldots, n\}$ and $j \geq 1$ with $i + j \leq n$ and $t^* > 0$ such that $X_{i+j}(t^*) > X_i(t^*)$, then there must be $0 \leq t_0 < t^*$ such that $X_{i+j}(t_0) = X_i(t_0)$. Let $F(t, Y) = BY[g(Yx(t)) - \frac{1}{x(t)} \sum_{k=1}^{n} f(x_k(t))]$. Using (3.4.10), we note that both X_{i+j} and X_i solve the initial value problem

$$\begin{cases} \dot{Y} = F(t, Y), \\ Y(t_0) = X_{i+j}(t_0) = X_i(t_0) \end{cases}$$

from which it follows that $X_{i+j}(t) = X_i(t)$ for all $t \geq t_0$, a contradiction to $X_{i+j}(t^*) > X_i(t^*)$. This completes the proof.

Note that the above proof shows that for a solution $(x_1(t), \ldots, x_n(t))$ of (3.4.1) with $0 < x_i(t) \leq B$ for $1 \leq i \leq n$, if $X_i(t^*) = X_j(t^*)$ for some $t^* \geq 0$ and some $i \neq j$, then $X_i(t) = X_j(t)$ for all $t \geq 0$.

With the above proposition, in the remaining part of this section, we will always label v_i so that $X_1(t) \leq \cdots \leq X_n(t)$ for all $t \geq 0$.

The following proposition describes an important condition under which limits of pattern variables always exist.

Proposition 3.4.5. *Let all x_i, $1 \leq i \leq n$, vary in a region I where $g(w)$ is monotonic. Then all the limits $Q_i = \lim_{t \to \infty} X_i(t)$ exist.*

Proof. Recall that $M(t) = \max\{X_i(t); i = 1, \ldots, n\} = X_n(t)$ and $m(t) = \min\{X_i(t); i = 1, \ldots, n\} = X_1(t)$. Using equation (3.4.12), we have

$$\dot{M}(t) \geq 0, \quad \dot{m}(t) \leq 0 \quad \text{for } t \geq 0, \text{ if } g \text{ is increasing on } I$$

or

$$\dot{M}(t) \leq 0, \quad \dot{m}(t) \geq 0 \quad \text{for } t \geq 0, \text{ if } g \text{ is decreasing on } I.$$

In both cases, Q_1 and Q_n exist and $Q_1 + Q_n > 0$ (More precisely, $Q_n > 0$ if g is increasing and $Q_1 > 0$ if g is decreasing).

We will consider the case where g is increasing, the case where g is decreasing can be dealt with similarly. We first show that Q_{n-1} exists. Integrating (3.4.11) at $i = n$ from $t = S$ to $t = T$, we obtain

$$X_n(T) - X_n(S) = B \int_S^T X_n(s) \sum_{k=1}^{n} X_k(s)[g(x_n(s)) - g(x_k(s))] \, ds. \qquad (3.4.13)$$

Letting $T \to \infty$ and using the fact that $X_n(t) \geq X_n(0)$ for $t \geq 0$, we get

$$Q_n - X_n(S) \geq X_n(0)h_{n,k}(S), \qquad (3.4.14)$$

where

$$h_{m,k}(s) = B \int_s^{\infty} X_k(v)|g(x_m(v)) - g(x_k(v))| \, dv, \quad 1 \leq m, k \leq n. \qquad (3.4.15)$$

Letting $S \to \infty$ in (3.4.14), we get

$$\lim_{S \to \infty} h_{n,k}(S) = 0. \qquad (3.4.16)$$

Also, using (3.4.11) for X_{n-1} and integrating over $[S, T] \subseteq [0, \infty)$ lead to

$$|X_{n-1}(T) - X_{n-1}(S)| \qquad (3.4.17)$$

$$\leq B \int_S^T \left[\sum_{k=1}^{n-1} X_k(s)|g(x_{n-1}(s)) - g(x_k(s))| + X_{n-1}(s)|g(x_n(s)) - g(x_{n-1}(s))| \right] ds.$$

Note that the definition of $h_{m,k}$, the monotonicity of g and the order-preserving property in Proposition 3.4.3 combined imply that

$$h_{n,k}(S) \geq h_{n-1,k}(S), \quad k = 1, 2, \ldots, n-1 \qquad (3.3.18)$$

from which and (3.4.16) it follows that

$$\lim_{S \to \infty} h_{n-1,k}(S) = 0, \quad k = 1, 2, \ldots, n-1. \tag{3.4.19}$$

Therefore, by (3.4.17), we conclude that Q_{n-1} exists.

Similarly, we can use the existence of Q_n and Q_{n-1} to prove the existence of Q_{n-2} by showing that $\lim_{S \to \infty} h_{m,k}(S) = 0$ for $m = n, n-1$ and $k = 1, \ldots, n$ implies that $\lim_{S \to \infty} h_{n-2,k}(S) = 0$ for $k \neq n-2$. Repeating the above argument n times, we then conclude that all Q_i, $1 \leq i \leq n$, exist. This completes the proof.

We now discuss how particular choices of f determine the limiting distribution Q_i.

Theorem 3.4.6 (Fair Distribution). *Let f be linear, namely, $f(w) = Cw$ for some $C > 0$. Then $Q_i = X_i(t) = X_i(0)$ for all $t \geq 0$. Denoted by $D = BC - A$. If $D > 0$, then the reverberation is persistent, $\lim_{t \to \infty} x_i(t) = X_i(0)C^{-1}D$ and $\lim_{t \to \infty} x(t) = C^{-1}D$ (therefore, reverberation normalization occurs). If $D \leq 0$, then the reverberation is transient.*

The above result shows that for a given linear f, if a pattern can reverberate persistently, then even small values of noise will reverberate albeit with small relative weight in the presence of large signals. This can be a liability in such systems, since in the absence of signals, noise will be amplified, and will receive a large relative weight.

Proof. The fact that $Q_i = X_i(t) = X_i(0)$ is an immediate consequence of equation (3.4.11), since $g(w) = \frac{f(w)}{w} = C$. Thus, $x_i(t) = X_i(0)x(t)$ and $G(x_1, \ldots, x_n) = C$. So by (3.4.9) we have

$$\dot{x} = -Ax + (B - x)Cx = [-A + BC - Cx]x. \tag{3.4.20}$$

Equation (3.4.20) can be solved by the method of separation to obtain

$$x_i(t) = \frac{x_i(0)e^{Dt}}{1 + x(0)CD^{-1}(e^{Dt} - 1)} \quad \text{if } D \neq 0,$$

$$x_i(t) = \frac{x_i(0)}{1 + x(0)Ct} \quad \text{if } D = 0,$$

from which all conclusions follow.

The following theorems show that if f grows faster than linearly, then noise can dissipate, and large values can quench small values before they amplified and maintained; but if f increases slower than linearly, then the opposite tendency occurs and the initial distribution is uniformized.

Theorem 3.4.7 (0-1 Distribution). *Let $f(w) = wg(w)$, where $g(w)$ is continuous, nonnegative and strictly increasing. If $M(0) = m(0) = \frac{1}{n}$, then $M(t) = m(t) = \frac{1}{n}$ for all $t \geq 0$. Otherwise, $M(t)$ is strictly increasing and $m(t)$ is decreasing (contour enhancement). Suppose, moreover, that the reverberation is persistent, then the limiting distribution is 0-1 or locally uniform, and satisfies $Q_1 = Q_2 = \cdots = Q_k = 0$ and $Q_{k+1} = \cdots = Q_n = (n-k)^{-1}$, where $X_k(0) < M(0) = X_{k+1}(0)$. Finally, the reverberation is persistent if $x(0) \geq \hat{x}$ with $\hat{x}$ being a root (if exists) of*

$$(n-k)M(0)g(M(0)x) = \frac{A}{B-x}.$$

Theorem 3.4.8 (Uniform Distribution). *Let $f(w) = wg(w)$, where $g(w)$ is continuous, nonnegative and strictly decreasing. Then the function $M(t)$ is decreasing and $m(t)$ is increasing (uniformization). Further, if the reverberation is persistent then $Q_i = \frac{1}{n}$ for $1 \leq i \leq n$. Finally, if in addition $g(0) > A/B$, then the reverberation is persistent.*

We note that if $f''(x) > 0$ for $x > 0$ then g is increasing and that if $f''(x) < 0$ for $x < 0$ then g is decreasing.

Proof of Theorem 3.4.7. If $M(0) = m(0) = \frac{1}{n}$, then $X_1(0) = X_2(0) = \cdots = X_n(0) = \frac{1}{n}$ from which and the remark following the proof of the order-preserving property it follows that $X_1(t) = \cdots = X_n(t) = \frac{1}{n}$ for all $t \geq 0$.

Suppose that $M(0) > m(0)$. If $M(t_0) = X_j(t_0)$, then $\dot{M}(t_0) \geq \dot{X}_j(t_0) > 0$ by (3.4.11). This shows that $M(t)$ is strictly monotonically increasing. Similarly, we can show that $m(t)$ is monotonically decreasing.

We now show the limiting distribution is 0-1, given a persistent reverberation, in the special case where $X_n(0) > X_{n-1}(0)$. The proof for locally uniform distribution in a general case is essentially the same.

Note that

$$X_i g(x_i) - X_j g(x_j) = (X_i - X_j)g(x_i) + X_j[g(x_i) - g(x_j)]$$

from which and (3.4.10) it follows that

$$\frac{d}{dt}[X_i - X_j] = BX_i[g(x_i) - G] - BX_j[g(x_j) - G]$$
$$= B(X_i - X_j)g(x_i) + BX_j[g(x_i) - g(x_j)] + BG(X_j - X_i).$$

Using (3.4.7), we get

$$\frac{d}{dt}[X_i - X_j] = B(X_i - X_j)\sum_{k=1}^{n} X_k[g(x_i) - g(x_k)] + BX_j[g(x_i) - g(x_j)]. \quad (3.4.21)$$

In particular, under the condition that $X_n(0) > X_{n-1}(0)$, we have

$$\frac{d}{dt}[X_n(t) - X_j(t)] \geq BX_j(t)[g(x_n(t)) - g(x_j(t))], \quad j \neq n. \quad (3.4.22)$$

Since g is strictly increasing and $X_n(t) \geq X_j(t)$ for $t \geq 0$, by (3.4.22) we have $X_n(t) - X_j(t) \geq X_n(0) - X_j(0) > 0$ for $j \neq n$ and $t \geq 0$. Also, since x varies in a positive closed interval, we can find a constant $\delta > 0$ such that for any $j \neq n$

$$g(x_n(t)) - g(x_j(t)) = g(X_n(t)x(t)) - g(X_j(t)x(t)) \geq \delta \text{ for all } t \geq 0.$$

Consequently,

$$\frac{d}{dt}[X_n(t) - X_j(t)] \geq \delta B X_j, \quad j \neq n.$$

Integrating this inequality from $t = 0$ to ∞ and using the fact that all $X_i \in [0, 1]$ yield

$$\infty > (\delta B)^{-1} \geq \int_0^\infty X_j(s)\, ds, \quad j \neq n,$$

from which and from the existence of $Q_j \in [0, 1]$ it follows that $Q_j = 0$ for $j \neq n$. Thus $Q_n = 1$.

We now prove the persistence in the case where $x(0) \geq \hat{x}$. Note that if $X_k(0) < M(0) = X_{k+1}(0)$, then $X_n(t) = \cdots = X_{k+1}(t) = M(t) > X_k(t)$ and $\dot{M}(t) > 0$. In particular, for $t > 0$

$$G \geq \sum_{i=k+1}^n X_i g(X_i x) = (n-k)M(t)g(M(t)x) > (n-k)M(0)g(M(0)x).$$

Thus, using (3.4.9), we have that when $x(t_0) = \hat{x}$ for some $t_0 > 0$, then

$$\dot{x}(t_0) = -A\hat{x} + (B - \hat{x})\hat{x}G(x_1(t_0), \ldots, x_n(t_0))$$
$$> -A\hat{x} + (B - \hat{x})\hat{x}(n-k)M(0)g(M(0)\hat{x}) = 0.$$

This shows that if $x(0) \geq \hat{x}$ then $x(t) \geq \hat{x} > 0$ for all $t \geq 0$, which proves the persistence. This completes the proof of Theorem 3.4.7.

We point out that if $g(0) \geq A/B$ and g is concave down then the reverberation is also persistent. See Grossberg [1970a].

Proof of Theorem 3.4.8. By Equation (3.4.11) and Proposition 3.4.4, $X_n(t) = M(t)$ is decreasing and $X_1(t) = m(t)$ is increasing. Thus $M(\infty) = \lim_{t\to\infty} M(t)$ and $m(\infty) = \lim_{t\to\infty} m(t)$ exist. We want to show if the reverberation is persistent then $M(\infty) = m(\infty)$, from which $Q_i = \frac{1}{n}$ for all i follows.

If $X_1(0) = X_n(0)$, then $X_1(0) = X_2(0) = \cdots = X_n(0)$. Therefore, $X_1(t) = X_2(t) = \cdots = X_n(t)$ for $t \geq 0$ from which $M(\infty) = m(\infty)$ follows. So we need only consider the case where $X_n(0) > X_1(0)$. By (3.4.21) and the decreasing property of g, we have

$$\frac{d}{dt}[X_n(t) - X_1(t)] \leq B X_1(t)[g(x_n(t)) - g(x_1(t))].$$

Note that

$$BX_1(t) \geq BX_1(0) > 0.$$

Therefore,

$$
\begin{aligned}
\frac{d}{dt}[X_n(t) - X_1(t)] &\leq -BX_1(0)[g(x_1(t)) - g(x_n(t))] \\
&= -BX_1(0)[g(X_1(t)x(t)) - g(X_n(t)x(t))] \\
&\leq -BX_1(0)[g(m(\infty)x(t)) - g(M(\infty)x(t))].
\end{aligned}
$$

If, by way of contradiction, $M(\infty) > m(\infty)$, then since g is strictly decreasing and since x varies in a positive closed interval, there exists $\delta > 0$ such that

$$\frac{d}{dt}[X_n(t) - X_1(t)] \leq -\delta < 0$$

for all $t \geq 0$. This yields a contradiction since both X_n and X_1 are bounded. Hence $M(\infty) = m(\infty)$.

We now show that if $g(0) > \frac{A}{B}$ then the reverberation is persistent. Using equation (3.4.9) and the fact that $Q_i = \lim_{t \to \infty} X_i(t)$ exists, we find that

$$\dot{x}(t) = -Ax(t) + (B - x(t))x(t) \sum_{k=1}^{n} Q_k g(Q_k x(t)) + \varepsilon(t)x(t), \qquad (3.4.23)$$

where

$$\varepsilon(t) \to 0 \quad \text{as } t \to \infty. \qquad (3.4.24)$$

More precisely, we have

$$\dot{x}(t) = \left[\left(\sum_{k=1}^{n} Q_k g(Q_k x(t)) - \frac{A}{B - x(t)} \right)(B - x(t)) + \varepsilon(t) \right] x(t). \qquad (3.4.25)$$

Since $g(0) > \frac{A}{B}$ and $\varepsilon(t) \to 0$ as $t \to \infty$, there exist $\delta_0 > 0$ and $T > 0$ so that

$$\left[\sum_{k=1}^{n} Q_k g(Q_k x) - \frac{A}{B - x} \right](B - x) \geq \frac{B}{2}\left[g(0) - \frac{A}{B} \right] \quad \text{if } 0 < x \leq \delta_0$$

and

$$|\varepsilon(t)| < \frac{B}{4}\left[g(0) - \frac{A}{B} \right] \quad \text{if } t \geq T.$$

We now claim that there exists $t^* \geq T$ so that

$$x(t^*) \geq \delta_0. \qquad (3.4.26)$$

Otherwise, we have from (3.4.25)

$$\dot{x}(t) \geq \frac{B}{4}\left[g(0) - \frac{A}{B}\right]x(t)$$

from which it follows that $x(t) \to \infty$ as $t \to \infty$, a contradiction. Therefore, $x(t^*) \geq \delta_0$ for some $t^* \geq T$. We then conclude that $x(t) \geq \delta_0$ for all $t \geq t^*$, since if $x(\tilde{t}) = \delta_0$ at $\tilde{t} \geq t^*$ then by (3.4.25), $\dot{x}(\tilde{t}) \geq \frac{B}{4}\left[g(0) - \frac{A}{B}\right]\delta_0 > 0$. Consequently, we have $x(t) \geq \delta_0$ for all $t \geq t^*$. This shows the reverberation is persistent and completes the proof.

Recall that x is bounded. Therefore, equation (3.4.23) implies that the "limiting equation" of (3.4.23) is given by

$$\dot{x} = -Ax + (B - x)x \sum_{k=1}^{n} Q_k g(Q_k x). \tag{3.4.27}$$

This is an autonomous ordinary differential equation. In the literature, equation (3.4.23) is said to be *asymptotically autonomous*. There is a very close relation between the limiting behaviors of solutions of (3.4.23) and those of its limiting equation (3.4.27). See Marcus [1956] and LaSalle [1976] for some classical work and Mischaikow, Smith and Thieme [1995] for some recent development regarding asymptotically autonomous differential equations. Applying these general results, one can show that in Theorem 3.4.8 $g(0) > \frac{A}{B}$ implies that $\lim_{t\to\infty} x(t) = E^*$, where E^* is the unique positive solution of $g\left(\frac{x}{n}\right) = \frac{A}{B-x}$. Consequently, $g(0) > \frac{A}{B}$ also implies that reverberation normalization occurs.

Figure 3.4.2 summarizes the impact of the choice of signal function on the signal enhancement and noise suppression in the on-center off-surround network.

If f is linear and the reverberation is persistent, the network can preserve an arbitrary pattern indefinitely. Unfortunately, the signal function amplifies noise in the absence of signals as vigorously as it amplifies signals. In the case where the signal function grows faster than a linear function and the reverberation is persistent, the limiting distribution exhibits an extreme form of contour enhancement and quenching whenever the initial pattern $X_i(0)$ is nonuniform: all pattern variables such that $X_i(0) < M(0)$ satisfy $Q_i \equiv 0$, while the maximal pattern weights $(X_i(0) = M(0))$ receive all the weight asymptotically. If a sufficiently energetic pattern is imposed upon noise, then the highest peaks of the pattern actively suppress both the noise and lesser pattern weights. If the signal function grows slower than linearly and the reverberation is persistent, the limiting distribution is uniform. Pattern uniformization can have unfortunate consequences in the presence of noise, since all states which receive either external inputs or random noise will asymptotically have equal importance.

Theorems 3.4.6–3.4.8 suggest how to construct functions f that will combine 0-1, fair and uniform tendencies. For example, define a continuous and positive $g(w)$

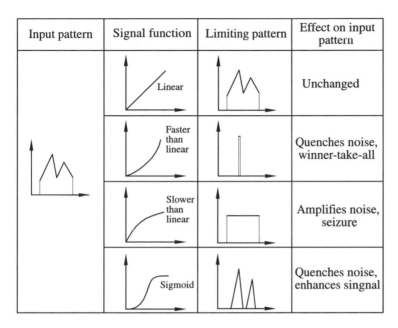

Figure 3.4.2. Signal enhancement and noise suppression in a feedback network for different choices of signal functions.

that is strictly increasing at all small values of w, and is strictly decreasing at large values of w. Theorems 3.4.7 and 3.4.8 suggest that 0-1 and uniform tendencies will be included in this way. A "fair" intermediate region can be constructed by choosing $g(w)$ constant (or, for all practical purposes, approximately so) between its increasing and decreasing values, since then $f(w)$ is linear in this range. See Figure 3.4.3.

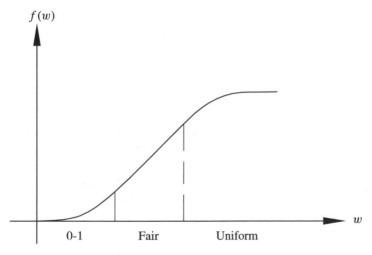

Figure 3.4.3. A sigmoid signal function which combines 0-1, fair and uniform tendencies.

The influence of these $f(w)$'s on the limiting distribution depends on particular choices of the parameters $x^{(0)}$, $x^{(1)}$ and $x^{(2)}$ defined respectively by

$$g(x^{(0)}) = g(B),\qquad\qquad(3.4.28)$$

$$x^{(1)} = \min\{w; g(w) = C\},\qquad\qquad(3.4.29)$$

$$x^{(2)} = \max\{w; g(w) = C\}\qquad\qquad(3.4.30)$$

shown in Figure 3.4.4.

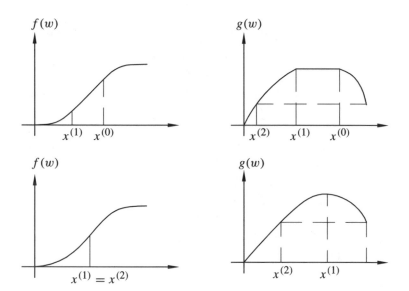

Figure 3.4.4. Important parameters in a sigmoidal signal function.

By constraining $x^{(0)}$, $x^{(1)}$ and $x^{(2)}$, we can achieve suitable subclasses of limiting possibilities. These constraints will be relevant to the observation that the slope of many realistic signal functions $f(w)$, such as sigmoid functions, eventually becomes horizontal, since the cells in a population have maximum response rates and other finite properties. The above results show that flattening of $f(w)$ can yield a uniform distribution. In the presence of noise, a uniformizing $f(w)$ imparts equal weight to essentially all states v_i, whether they are excited by signals or not, after sufficient reverberation has taken place. The flattening of $f(w)$ can thus be disadvantageous to effective signal processing.

The following result shows that suitable choices of $x^{(0)}$, $x^{(1)}$ and $x^{(2)}$ can prevent uniformization even if maximum value B of $x(t)$ exceeds $x^{(2)}$ and therefore lies in the uniformizing range.

Theorem 3.4.9. *Let $f(w) = wg(w)$, where $g(w)$ is continuous and nonnegative and*

$$g(w) \begin{cases} \text{is increasing for } 0 \leq w \leq x^{(1)}; \\ \text{equal to } C \text{ for } x^{(1)} \leq w \leq x^{(2)}; \\ \text{is decreasing for } x^{(2)} \leq w \leq B. \end{cases}$$

Then we have the following conclusions.

(I) **Fair:** *If*

$$X_1(0) \min\{B - AC^{-1}, x(0)\} \geq x^{(1)} \tag{3.4.31}$$

and

$$X_n(0) \max\{B - AC^{-1}, x(0)\} \leq x^{(2)}, \tag{3.4.32}$$

then $Q_i = X_i(t) = X_i(0)$, $i = 1, \ldots, n$, and $x(t)$ approaches $E - B - AC^{-1}$ monotonically as $t \to \infty$.

(II) **Fair, or Contour Enhancing and Quenching:** *Let*

$$x^{(0)} + x^{(2)} \geq \max\{B - AC^{-1}, x(0)\}. \tag{3.4.33}$$

Then all Q_i exist, $M(t)$ is increasing and $m(t)$ is decreasing. The limiting distribution is either fair, 0-1, locally uniform (only if several $X_i(0) = M(0)$), or contour enhancing and quenching; no uniformization occurs. If, moreover, for some $L < n$,

$$X_L(0) \min\{B - \frac{A}{\sum_{i=L}^{n} X_i(0)C}, x(0)\} \geq x^{(1)}, \tag{3.4.34}$$

then $\frac{dX_1(t)}{dt} \geq 0$ and $\frac{d}{dt}[X_i(t)X_j^{-1}(t)] = 0$ for $t \geq 0$ and $i, j \geq L$ (so that contour enhancement occurs). If

$$X_1(0) \min\{B - AC^{-1}, x(0)\} \geq x^{(1)}, \tag{3.4.35}$$

then $Q_i = X_i(t) = X_i(0)$ for $t \geq 0$ and $i = 1, \ldots, n$. If, however, the reverberation is persistent and

$$B - AC^{-1} < Nx^{(1)} \tag{3.4.36}$$

with $1 < N \leq n$ and $X_{n-N+1}(0) < X_n(0)$, then $Q_1 = \cdots = Q_{n-N+1} = 0$ (so the quenching occurs). If

$$X_1(0)(B - AC^{-1}) \leq x^{(0)}, \tag{3.4.37}$$

and if the reverberation is persistent and if $X_1(0) < X_n(0)$, then $Q_1 = 0$. If

$$X_i(t_i)(B - AC^{-1}) \leq x^{(0)} \tag{3.4.38}$$

for a sufficient large $t = t_i$, then $Q_i = 0$ if $X_i(t_i) < X_n(t_i)$ and the reverberation is persistent. If $x^{(1)} = x^{(2)}$ and the reverberation is persistent, then the limiting distribution is 0-1 if $X_n(0) > X_{n-1}(0)$, and locally uniform otherwise.

(III) *Quenching:* If

$$X_1(0) \max\{B - AC^{-1}, x(0)\} \leq x^{(0)}, \qquad (3.4.39)$$

then $\frac{\mathrm{d}}{\mathrm{d}t}X_1(t) \leq 0$ for $t \geq 0$. If, moreover, $X_1(0) < X_n(0)$ and the reverberation is persistent, then $Q_1 = 0$. If (3.4.38) holds for t_i sufficiently large and $i < n$, then $\frac{\mathrm{d}}{\mathrm{d}t}X_i(t) \leq 0$ for $t \geq t_i$, and $Q_i = 0$ if the reverberation is persistent and $X_i(t_i) < X_n(t_i)$.

(IV) *Quenching:* If

$$(n - 1)x^{(0)} + x^{(2)} > \max\{B - AC^{-1}, x(0)\}, \qquad (3.4.40)$$

then $\frac{\mathrm{d}}{\mathrm{d}t}X_1(t) \leq 0$ for $t \geq 0$. If, moreover, $X_1(0) < X_n(0)$ and the reverberation is persistent, then $Q_1 = 0$.

(V) *Uniformizing:* If for some K, $0 < K < n$,

$$(n - K + 1)x^{(2)} \geq \max\{B - AC^{-1}, x(0)\} \qquad (3.4.41)$$

and

$$X_1(0) \min\{B - \frac{A}{\sum_{i=1}^{K} X_i(0)C}, x(0)\} \geq x^{(1)}, \qquad (3.4.42)$$

then all Q_i exist with $Q_1 < 1/n < Q_n$, even though $\frac{\mathrm{d}}{\mathrm{d}t}X_i(t) \geq 0$ and $\frac{\mathrm{d}}{\mathrm{d}t}X_n(t) \leq 0$ for $t \geq 0$.

The proof of Theorem 3.4.9 is quite technical and we refer to Grossberg [1970a] for more details. Note that although Theorem 3.4.9 provides practical constraints on $x^{(0)}$, $x^{(1)}$, $x^{(2)}$ that guarantee functionally useful behavior, it has not yet been proved in general that all Q_i exist in the absence of constraints. Such a theorem would be of particular interest in pathological conditions where $x^{(0)}$, $x^{(1)}$ and $x^{(2)}$ might deviate from normal values. Can sustained oscillations occur in pathological cases? In Grossberg [1970a], it is shown that interaction between x and $(X_1, \ldots, X_n)$ can produce oscillations, but whether these oscillations always dissipate remains to be proved.

In summary, we have noticed that recurrent on-center off-surround anatomies permit contour enhancement of sensory information (see Ratcliff [1965]), allow effective processing of arbitrary input patterns with fluctuating intensities, and are capable of short term memory in the sense that they can reverberate a pattern of activity distributed over cell populations for an indefinite interval of time.

3.5   Determining synaptic weights

It is important to determine the synaptic coupling coefficients for neural networks with specified interconnection topology. In this section, we follow the presentation of

Harvey [1994] and describe a method based on genetic algorithm (GA), a procedure suggested by natural heredity and evolution for efficiently searching over a parameter space.

We consider a special example to design a neural network for feature detectors. A common design is to compute the output by convolving the input pattern with a kernel matrix. The designer selects the kernel matrix so that the output measures, say, the orientation of an edge in the input pattern. The response mimics the orientation responses to illumination contrast of simple cells in the visual cortex.

Let us start with the formulation of the model equations. Suppose a set of neurons $\{v_i\}$ form a feature detector module. We characterize the ith neuron by its activation level x_i and by its connections with other neurons $\{w_{ji}\}$. This is a hardwiring problem with specified connection topology, so we assume the synaptic coupling coefficients are constants (to be determined), and thus we will not consider the evolution of LTM traces.

For short-term memory (STM), we assume an equation for v_i of the form

$$\frac{dx_i}{dt} = -\alpha x_i + \sum_j w_{ij} f(x_j) + I_i, \qquad (3.5.1)$$

where

x_i = activation of the ith neuron,

w_{ij} = long-term memory (LTM) trace from the jth neuron to the ith neuron,

I_i = external input to the ith neuron,

f = a nonlinear signal function,

α = relaxation time constant parameter.

Our goal is to determine weights $\{w_{ij}\}$ to satisfy prescribed input/output relations using the GA. We demonstrate the GA by considering a simple task of the design of a feature detector module that gives a single output for some spatial activation pattern defined on an array of input neurons. The output could show, say, the angular orientation of the pattern.

We assume input patterns defined on M input neurons, N hidden (internal) neurons, and a single output neuron. Thus, each input pattern is characterized by the activation level of a single output neuron. We assume no feedback from the hidden neurons or from the output neuron to the input neurons, but we allow feedback among the hidden neurons and assume direct connections from the input neurons to the output neuron. Therefore, we can write a set of STM equations from (3.5.1) as follows.

- **Input Neurons**

$$\frac{dx_i}{dt} = -\alpha x_i + I_i, \quad i = 1, \ldots, M. \qquad (3.5.2)$$

- **Hidden Neurons**

$$\frac{dx_i}{dt} = -\alpha x_i + \sum_{j=1}^{M+N} w_{ij} f(x_j), \quad i = 1, \ldots, N. \tag{3.5.3}$$

- **Output Neuron**

$$\frac{dx_0}{dt} = -\alpha x_0 + \sum_{j=1}^{M+N} w_{0j} f(x_j). \tag{3.5.4}$$

Note that the STM equations of the hidden neurons can be rearranged as

$$\frac{dx_i}{dt} = -\alpha x_i + \sum_{j=1}^{N} w_{ij} f(x_j) + \sum_{k=1}^{M} w'_{ik} f(x_k), \quad i = 1, \ldots, N, \tag{3.5.5}$$

where

$$w_{ij} = \text{LTM trace of the hidden neurons},$$

$$w'_{ik} = \text{LTM trace from the input neurons to the hidden neurons.}$$

The network described by (3.5.2)–(3.5.5) is a special case of the so-called "a cascade of neural nets" whose convergence was investigated in Hirsch [1989] and Smith [1991].

Assume $\alpha = 1$ (equivalent to rescaling the other variables). Then, the input neuron activations approach the external inputs, that is, $x_k \to I_k, k = 1, \ldots, M$ as $t \to \infty$.

Assume every solution of (3.5.2)–(3.5.5) is convergent to the set of equilibria. Then, in the steady state the hidden neuron activations become

$$x_i = \sum_{j=1}^{N} w_{ij} f(x_j) + \sum_{k=1}^{M} w'_{ik} f(I_k), \quad i = 1, \ldots, N. \tag{3.5.6}$$

Let $x = (x_1, \ldots, x_N)^T$, $A = (w_{ij})_{1 \le i, j \le N}$, $B = (w'_{ik})_{1 \le i \le N, 1 \le k \le M}$ and $f(I) = (f(I_1), \ldots, f(I_N))^T$. Then the steady-state hidden neuron equation (3.5.6) in matrix notation becomes

$$x = A f(x) + B f(I). \tag{3.5.7}$$

Let the steady-state output be Z, that is, $Z = \lim_{t \to \infty} x_0(t)$. Then (3.5.4) gives

$$Z = \sum_{j=1}^{N} w_{0j} f(x_j) + \sum_{k=1}^{M} w'_{ok} f(I_k), \tag{3.5.8}$$

which in matrix form is

$$Z = C f(x) + D f(I), \tag{3.5.9}$$

where $C = (w_{01}, \ldots, w_{0N})$ and $D = (w'_{01}, \ldots, w'_{0M})$.

Matrix A gives the connections among N hidden neurons. Matrix B gives the connections from M input neurons to the hidden neurons. Matrix C gives connections from the hidden neurons to a single output neuron. Matrix D gives connections from the inputs to the output. Thus, four matrices $\{A, B, C, D\}$ describe the neural network, and the task of designing a neural network with given input-output characteristics becomes the task to determine matrices $\{A, B, C, D\}$ for given input-output relations.

To illustrate the design method, we assume the neural network is to model — crudely — biological feature detectors such as those found in the human primary visual cortex. So, we follow the work of Crick and Asanuma [1988] based on some experimental findings and formulate the following assumptions for mimicking biological neural networks:

(H1) Each neuron is excitatory or inhibitory and cannot be both types.

(H2) A neuron cannot excite or inhibit itself by axon signals.

(H3) The neural networks are in an on-center off-surround architecture.

With these assumptions, the system matrices for on-center off-surround neural networks have the following properties:

(P1) A has zero diagonal components.

(P2) A and C have negative or zero elements.

(P3) B and D have positive or zero elements.

(P4) The columns of A give the lateral inhibition to other hidden neurons.

We now describe a design procedure based on *generic algorithm* (GA). The GA is a search procedure, inspired by evolution and heredity, for finding high performance structures in a complex parameter domain. For neural network design purpose, the GA is a search method for finding a good set of weights in a high-dimensional nonlinear weight space, degenerate forms of the GA are the well-known gradient-descent techniques, including backward propagation.

Our computational steps for the design procedure based on GA are as follows:

1. Randomly choose a beginning matrix set $\{A, B, C, D\}$ with the properties (P1)–(P4).

2. Copy the parent set $\{A, B, C, D\}$. In each copy, randomly select columns of the parent. Randomly change the elements of the selected columns subject to the constraints.

3. Define input training patterns. For each training pattern and each copy, solve for the output. (*Note:* A solution may not always exist — see below.)

4. Select the survivor as the parent for the next generation, repeat Steps 2 to 4.

5. Continue until the payoff criterion is met. The surviving system $\{A, B, C, D\}$ describes the neural network.

Figure 3.5.1 shows a flow diagram of the procedure. Clearly, it is essential to define a good payoff function. For illustration, we choose a metric that measures the distance

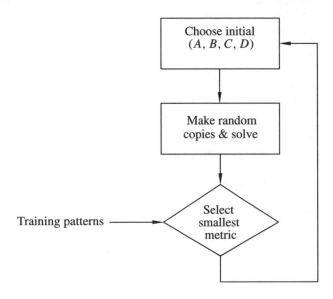

Figure 3.5.1. Computational flow diagram of the design method.

(error) of the output from a desired output for each training pattern. Many metrics — or more precisely pseudometrics — are applicable. A practical metric specifies that, for each input training pattern, the output response lies in a band of upper and lower thresholds. The designer specifies the thresholds. We now formulate the HI-LO metric as follows: For N_p training patterns, the metric d is

$$d = \sum_{i=1}^{N_p} s(LB_i \leq Z_i \leq UB_i), \tag{3.5.10}$$

where

$$Z_i = \text{output for the } i\text{th input training pattern,}$$
$$LB_i = \text{lower bound for the } i\text{th input pattern,}$$
$$UB_i = \text{upper bound for the } i\text{th input pattern,}$$

and

$$s(LB_i \leq Z_i \leq UB_i) = \begin{cases} 0 & \text{if } LB_i \leq Z_i \leq UB_i, \\ 1 & \text{otherwise.} \end{cases} \tag{3.5.11}$$

Consequently, the condition $d = N_p$ means the system $\{A, B, C, D\}$ satisfies none of the payoff criteria (maximum error), while the condition $d = 0$ means the system satisfies the selection criterion and that a solution has been reached. In practice, the metric d starts at N_p (or smaller) and monotonically decreases to zero.

To implement Step 3 of the algorithm, for each set $\{A, B, I\}$, we have to solve the equation (3.5.7). Three outcomes are possible when trying to solve (3.5.7): No solutions may exist, a single solution may exist, or multiple solutions may exist. In general, one can solve (3.5.7) by iteration if such solutions exist. Namely, we define

$$x_0 = Bf(I),$$
$$x_1 = Af(x_0) + x_0,$$

and in general,

$$x(n+1) = Af(x(n)) + Bf(I), \quad n \geq 1, \tag{3.5.12}$$

where

$$x_1 = \text{first trial solution},$$
$$x(n) = \text{current trial solution for } x,$$
$$x(n+1) = \text{next trial solution for } x.$$

The recursion is continued until $x(n+1) - x(n)$ becomes sufficiently small. The recursion converges to a solution if the operator defined by the right-hand of (3.5.7) is a contraction operator, a specific example will be given in the next chapter.

Let us demonstrate the above design by an example which is an on-center off-surround neural network sensitive to horizontal binary patterns on a square array of neurons that has a resolution of $45°$. Table 3.5.2 shows the assumed parameters for the GA design example. The example assumes an input pattern defined on 7×7 or 49 input neurons.

Input neurons (M)	7×7 (49)
Hidden neurons (N)	25
Copies per generation	10
Fraction of weights changed per copy	1/3
Search range	$0, \pm 1, \pm 2, \dots, \pm 10$
Hign band	50 to 100
Low bound	-100 to 10

Table 3.5.2. Input parameters for designing a horizontal feature detector using an on-center off-surround neural network.

Starting with a random set $\{A, B, C, D\}$ for the first generation, at each generation ten copies are made of the parent set. For each copy, a third of the matrix elements (weights) are randomly changed by selecting integer values over the range -10 to 10, subject to the constraints given in (P1)(P4). Assume a HI-LO metric for the training patterns with some of the outputs in a high band and others in a low band. That is, the desired output response to the high (horizontal) training patterns is 50 to 100. The desired response to the low (nonhorizontal) training patterns is -100 to 10.

Row-by-row scanning produces the input vector I for each training pattern. That is, I_1 to I_7 are the first row, I_8 to I_{14} are the second row, and I_{43} to I_{49} are the seventh row. The example determines $1924 = 49 \times 25 + 25 \times 25 + 49 + 25$ coefficients for this neural network. For simplicity, assume the signal function is a unit step function. Compute the HI-LO metrics of the ten copies and compare with the metric of the parent set. If an offspring metric is below the parent metric, the offspring replaces the parent set for the next generation.

Figure 3.5.3 shows 12 training patterns. For a horizontal detector, assume the desired responses are high (50 to 100) to the horizontal patterns and are low (-100 to 10) to the others.

0	0	0	0	0	0	0
0	0	0	0	0	0	0
0	0	0	0	0	0	0
1	1	1	1	1	1	1
0	0	0	0	0	0	0
0	0	0	0	0	0	0
0	0	0	0	0	0	0

0	0	0	0	0	0	1
0	0	0	0	0	1	0
0	0	0	0	1	0	0
0	0	0	1	0	0	0
0	0	1	0	0	0	0
0	1	0	0	0	0	0
1	0	0	0	0	0	0

0	0	0	0	0	0	0
0	0	0	0	0	0	0
0	0	0	0	0	0	0
0	1	1	1	1	1	0
0	0	0	0	0	0	0
0	0	0	0	0	0	0
0	0	0	0	0	0	0

0	0	0	0	0	0	0
0	0	0	0	0	1	0
0	0	0	0	1	0	0
0	0	0	1	0	0	0
0	0	1	0	0	0	0
0	1	0	0	0	0	0
0	0	0	0	0	0	0

0	0	0	0	0	0	0
0	0	0	0	0	0	0
0	0	0	0	0	0	0
0	0	1	1	1	0	0
0	0	0	0	0	0	0
0	0	0	0	0	0	0
0	0	0	0	0	0	0

0	0	0	0	0	0	0
0	0	0	0	0	0	0
0	0	0	0	1	0	0
0	0	0	1	0	0	0
0	0	1	0	0	0	0
0	0	0	0	0	0	0
0	0	0	0	0	0	0

Figure 3.5.3 (a): Horizontal 45 degrees

```
0  0  0  1  0  0  0        1  0  0  0  0  0  0
0  0  0  1  0  0  0        0  1  0  0  0  0  0
0  0  0  1  0  0  0        0  0  1  0  0  0  0
0  0  0  1  0  0  0        0  0  0  1  0  0  0
0  0  0  1  0  0  0        0  0  0  0  1  0  0
0  0  0  1  0  0  0        0  0  0  0  0  1  0
0  0  0  1  0  0  0        0  0  0  0  0  0  1

0  0  0  0  0  0  0        0  0  0  0  0  0  0
0  0  0  1  0  0  0        0  1  0  0  0  0  0
0  0  0  1  0  0  0        0  0  1  0  0  0  0
0  0  0  1  0  0  0        0  0  0  1  0  0  0
0  0  0  1  0  0  0        0  0  0  0  1  0  0
0  0  0  1  0  0  0        0  0  0  0  0  1  0
0  0  0  0  0  0  0        0  0  0  0  0  0  0

0  0  0  0  0  0  0        0  0  0  0  0  0  0
0  0  0  0  0  0  0        0  0  0  0  0  0  0
0  0  0  1  0  0  0        0  0  1  0  0  0  0
0  0  0  1  0  0  0        0  0  0  1  0  0  0
0  0  0  1  0  0  0        0  0  0  0  1  0  0
0  0  0  0  0  0  0        0  0  0  0  0  0  0
0  0  0  0  0  0  0        0  0  0  0  0  0  0
```

Figure 3.5.3 (b): Vertical 135 degrees

Figure 3.5.3. Binary training patterns for a design example of an ON-CTR/OFF-SUR architecture feature detector with an angular resolution of $45°$.

Figure 3.5.4 shows the history of the metric as the system evolves to a solution in about 600 generations. The metric started at $d = 10$ and ended at $d = 1$ at 600 generations. The one remaining error was a response above the high threshold, and so the run was stopped.

Appendix. Grossberg's Learning Theorem

We consider two types of network connections which are suitable for pattern learning. Namely, we consider the case where a finite number of cells $\mathcal{A}$ send axons to a finite number of cells $\mathcal{B}$, and either $\mathcal{A} = \mathcal{B}$ or $\mathcal{A} \cap \mathcal{B} = \emptyset$, and we discuss the interaction ("pairing") of cells $\mathcal{A}$ activated by any finite number of conditional stimulus with the cells $\mathcal{B}$ activated by any unconditional stimulus. In mathematical terminology, let K and J be finite (indices) sets of integers and assume either $J = K$ or $J \cap K = \emptyset$. Let $x_k : \mathbb{R} \to \mathbb{R}$, $x_j : \mathbb{R} \to \mathbb{R}$ and $Z_{kj} : \mathbb{R} \to \mathbb{R}$ be continuous functions. Let $W_{(-\infty, t]}$ denote the restriction of all (x_j, x_k, Z_{kj}) to the interval $(-\infty, t]$ (the history of (x_j, x_k, Z_{kj}) up to the moment t). Assume that $x_j(t)$, $x_k(t)$ and $Z_{kj}(t) \geq 0$ for all

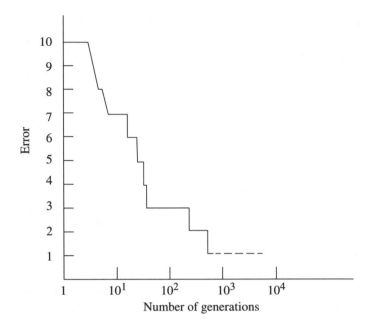

Figure 3.5.4. Time history of the error measuring derivation of the outputs of 12 training patterns from specific outputs.

$t \leq 0$ and $j \in J, k \in K$; and $x_j(t), x_k(t)$ and $Z_{kj}(t)$ are differentiable for $t > 0$, $j \in J, k \in K$ and assume that

$$\begin{cases} \dot{x}_j(t) = A(W_{(-\infty,t]}, t)x_j(t) + \sum_{k \in K} B_k(W_{(-\infty,t]}, t)Z_{kj}(t) + \theta_j I(t), \\ \dot{Z}_{kj}(t) = D_k(W_{(-\infty,t]}, t)Z_{kj}(t) + E_k(W_{(-\infty,t]}, t)x_j(t) \end{cases} \quad (\text{I}.1)$$

for $t > 0$, $j \in J$ and $k \in K$, where A, B_k, D_k and E_k are continuous functions of $t \geq 0$, and B_k and E_k are nonnegative. Note that x_k represents the output of cell $k \in K$ generated by conditional stimulus, such as x_0 in (3.1.1). We do not write equations for these variables.

An important constraint in the general model (I.1) and the simple model (3.1.1) is that the synaptic delays from a given cell for signal transfer to all the cells in a functionally coordinated unit depend only on the source cell. A natural question arises: how can different axons from the source cell have the same time lag if they have different lengths? The explanation, offered in Grossberg [1968a, 1971a], is as follows: Let axon length be proportional to axon diameter and then let signal velocity be proportional to axon diameter. The latter is often the case (Ruch, Patton, Woodbury and Towe [1961]) and the former is qualitatively true since longer axon of a given cell type are usually thicker.

The Grossberg's Learning Theorem can be stated as follows:

Theorem 3.5.1. *Assume that*

(i) x_j *and* Z_{kj} *are bounded functions of* $t \geq 0$ *for* $j \in J$, $k \in K$;

(ii) $\int_0^\infty I(v) \, dv = \infty$;

(iii) *there exist positive constants* T_1 *and* T_2 *such that for all* $T \geq 0$,

$$\int_T^{T+t} I(v) \left\{ \exp \left[\int_v^{T+t} A(W_{(-\infty,\xi]}, \xi) \, d\xi \right] \right\} dv \geq T_1 \quad \text{if } t \geq T_2;$$

(iv) $\int_0^\infty B_k(W_{(-\infty,v]}, v) \, dv = \infty$ *if* $\int_0^\infty E_k(W_{(-\infty,v]}, v) \, dv = \infty$.

Then $\lim_{t \to \infty} \frac{Z_{kj}(t)}{\sum_{m \in J} Z_{km}(t)} = P_{kj}$ *and* $\lim_{t \to \infty} \frac{x_j(t)}{\sum_{m \in J} x_m(t)} = Q_j$ *exist and* $Q_j = \theta_j$.

Moreover, if

(v) $\int_0^\infty E_k(W_{(-\infty,v]}, v) \, dv = \infty$,

then $P_{kj} = \theta_j$.

We refer to Grossberg [1969] for a rigorous proof of this theorem. (ii) and (iii) require that the UCS is presented sufficiently often, and conditions (iv) and (v) combined imply that each CS is presented sufficiently often and each CS and the UCS are practiced together sufficiently often. The learning theorem claims that a network subject to (i)–(v) can learn a spatial pattern.

When applying the above result to our model (3.1.1) for an outstar, we have $K = \{0\}$, $J = \{1, \ldots, n\}$, $A = -\alpha$, $\theta_j I(t) = I_j(t)$, $D_j = -\beta$,

$$B_k(W_{(-\infty,t]}, t) = f(x_0(t - \tau) - \theta),$$
$$E_k(W_{(-\infty,t]}, t) = g(x_0(t - \tau) - \theta)$$

with

$$x_0(t) = e^{-\alpha t} x_0(0) + \int_0^t e^{-\alpha(t-s)} I_0(s) \, ds.$$

Therefore, we have

Corollary 3.5.2. *Assume that*

(i) $x_i(0) \geq 0$ *and* $Z_i(0) \geq 0$ *for* $i = 1, \ldots, n$; $x_0(s)$ *is a continuous function of* $s \in [-\tau, 0]$;

(ii) $\int_0^\infty f(e^{-\alpha t} x_0(0) + \int_0^t e^{-\alpha(t-s)} I_0(s) \, ds - \theta) dt = \infty$ *if* $\int_0^\infty g(e^{-\alpha t} x_0(0) + \int_0^t e^{-\alpha(t-s)} I_0(s) \, ds - \theta) dt = \infty$;

(iii) *there exist positive constants K_1 and K_2 such that for all $T \geq 0$,*

$$\int_T^{T+t} e^{-\alpha(T+t-v)} I(v)\, dv \geq K_1 \quad \text{if } t \geq K_2;$$

(iv) $x_i(t)$ *and* $Z_i(t)$ *are bounded for* $t \geq 0$ *and* $1 \leq i \leq n$.

Then $\lim_{t\to\infty} \frac{Z_i(t)}{\sum_{j=1}^n Z_j(t)} = P_i$ *exists and* $\lim_{t\to\infty} \frac{x_i(t)}{\sum_{j=1}^n x_j(t)} = \theta_i$.

Moreover, if

(v) $\int_0^\infty g(e^{-\alpha t} x_0(0) + \int_0^t e^{-\alpha(t-s)} I_0(s)\, ds - \Gamma)\, dt = \infty$,

then $P_i = \theta_i$.

Condition (i) implies that all solutions of (3.1.1) remain nonnegative for all $t \geq 0$. Note that if $(I_0, I_1, \ldots, I_n)$ are all constants, then (ii), (iii) and (v) are satisfied if $\frac{I_0}{\alpha} > \Gamma$ and if $f(x) > 0$, $g(x) > 0$ for all $x > 0$. As discussed in Section 3.1, the condition $\alpha\beta > f\left(\frac{I_0}{\alpha} - \Gamma\right) g\left(\frac{I_0}{\alpha} - \Gamma\right)$ implies the boundedness of solutions. Finally, it is an easy exercise to verify that (iii) implies that $\int_0^\infty I(v) dv = \infty$.

Generalization to learning an arbitrary space-time pattern can be readily accomplished. See Grossberg [1970a]. See also Grossberg [1970b, 1971b] for applications to pattern discrimination.

In applications, the inputs are more complicated than constant functions, and verification of conditions (ii) and (v) can be very difficult. Similar models for prediction and learning were discussed in Grossberg [1967, 1968a–c].

Chapter 4

Content-addressable memory storage

In a dynamical system generated by a differential equation, an *equilibrium* represents a steady state of the system which does not change in time. A *(asymptotically) stable steady state* (equilibrium) is one which has a neighborhood such that the trajectory converges to the equilibrium if it initially starts from a point in such a neighborhood. See Figure 4.1 below.

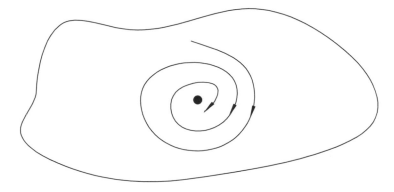

Figure 4.1. A stable steady state (equilibrium) and the limiting behavior of a trajectory starting from a small neighborhood of the steady state.

Content-addressable memory (CAM) is one of the simplest properties of a wide variety of neural network models. This property involves the change of the network with time where the motion of the activities (or states) of neurons describes the computation that the network is performing. In mathematical terms, the evolution of a network defines a dynamical system or a flow in a phase space. The network is developing stable steady states which are related to stored information. States near to the particular stable equilibrium point contain partial information about the memory, from an initial state of partial information about the memory, a final stable steady state with all the information of the memory can be found due to the aforementioned convergence. The memory is reached not by knowing an address, but rather by supplying in the initial state some subpart of the memory. Any subpart of adequate size will do —the memory is truly addressable by *content* rather than location. Of course, a network may contain many memories simultaneously, and this is reflected by the fact that the associated dynamical system exhibits coexistence of multiple stable equilibria.

From a mathematical perspective, the question of content-addressable memory in a neural network can be formulated as follows:

(a) Under what conditions does a network approach an equilibrium point in response to the arbitrary initial data?

(b) How many equilibria exist and which are stable?

(c) How does the system in response to a given initial data approach an equilibrium?

This is clearly related to the *global* analysis of the dynamics of the network and, in most cases, is extremely difficult.

In this chapter, we are going to introduce several important approaches towards the global analysis of a network of neurons. One approach which has been proved to be very effective is the use of a *Liapunov (or energy) function*. Another alternative is the powerful theory of *monotone dynamical systems* recently developed in association with biological competitive-cooperative systems.

4.1   Parallel memory storage by competitive networks

We are going to present a convergence theorem of Cohen and Grossberg [1983], based on the construction of a Liapunov (energy) function and the application of the so-called LaSalle Invariance Principle.

Let us start with a general autonomous system of ordinary differential equations

$$\frac{dx}{dt} = \dot{x} = f(x), \tag{4.1.1}$$

where $f : G \longrightarrow \mathbb{R}^n$ is a continuous mapping defined on an open subset G of $\mathbb{R}^n$, and we assume that for each $x_0 \in \mathbb{R}^n$, (4.1.1) has a unique solution, denoted by $\Phi(t, x_0)$, which satisfies $\Phi(0, x_0) = x_0$ and is defined for all $t \geq 0$. A point $y \in \mathbb{R}^n$ is said to be an ω-*limit point* of $x_0 \in \mathbb{R}^n$ if there exists a sequence $\{t_m\}_{m=1}^{\infty}$ such that $t_m \geq 0$, $t_m \to \infty$ and $\Phi(t_m, x_0) \to y$ as $m \to \infty$. The set of all ω-limit points of x_0 is called the ω-*limit set* of x_0 and is denoted by $\omega(x_0)$. Clearly, this contains all the information of the limiting behaviors of $\Phi(t, x_0)$ as $t \to \infty$. A set $H \subseteq G$ is said to be *positively invariant* if $x_0 \in H$ implies $\Phi(t, x_0) \in H$ for all $t \geq 0$; *invariant* if for each $x_0 \in H$, (4.1.1) has a unique solution $\Phi(t, x_0)$ defined for all $t \in \mathbb{R}$ and satisfying $\Phi(0, x_0) = x_0$, and $\Phi(t, x_0) \in H$ for all $t \in \mathbb{R}$. It is known that if the *trajectory* $\Phi(\cdot, x_0) : [0, \infty) \to G$ is bounded, then $\omega(x_0)$ is nonempty, compact, invariant and connected and dist$(\Phi(t, x_0), \omega(x_0)) \to 0$ as $t \to \infty$, where dist$(x, A) = \inf_{a \in A} ||x - a||$ for $x \in \mathbb{R}^n$ and for a subset $A \subseteq \mathbb{R}^n$. See Hale [1969].

Let $V : G \to \mathbb{R}$ be a given continuous mapping. Define

$$\dot{V}_{(4.1.1)}(x) = \liminf_{t \to 0^+} \frac{1}{t} [V(\Phi(t, x)) - V(x)] \tag{4.1.2}$$

for any $x \in G$. In many applications, V is C^1-smooth (that is, V has continuous partial derivatives on D). In this case,

$$\dot{V}_{(4.1.1)}(x) = (\text{grad } V(x), f(x)), \quad \text{grad } V(x) = \left(\frac{\partial V(x)}{\partial x_1}, \dots, \frac{\partial V(x)}{\partial x_n} \right) \quad (4.1.3)$$

and thus $\dot{V}_{(4.1.1)}(x)$ can be computed directly from the differential equation (4.1.1) without finding the solutions explicitly.

We say V is a *Liapunov function* of (4.1.1) on a subset $D \subseteq G$ if $\dot{V}_{(4.1.1)}(x) \le 0$ for all $x \in D$. Therefore, for a Liapunov function V of (4.1.1) on D and for a solution $\Phi(t, x) \in D$ for $t \ge 0$, $V(\Phi(t, x))$ is nonincreasing with respect to $t \in [0, \infty)$.[1] Relative to a Liapunov function V of (4.1.1) on D, we can introduce the set

$$E = \{x; \dot{V}_{(4.1.1)(x)} = 0, \ x \in \overline{D} \cap G\},$$
$$M = \text{ the largest invariant set in } E.$$

The following result, called *LaSalle's Invariance Principle*, provides an effective way to locate the ω-limit set of a bounded solution of (4.1.1) via a Liapunov function. We refer to LaSalle [1976] for more details.

Theorem 4.1.1. *Let $V : G \to \mathbb{R}$ be a Liapunov function of (4.1.1) on D and let $x(t) = \Phi(t, x_0)$ be a solution of (4.1.1) that is bounded and remains in D for all $t \ge 0$. Then there exists some $c \in \mathbb{R}$ such that $\omega(x_0) \cap G \subset M \cap V^{-1}(c)$.*

The key to the application of this result is the construction of a Liapunov function. Unfortunately, there is no general method for constructing Liapunov functions. A notable success is MacArthur [1969] where he constructed a Liapunov function for Gause–Lotka–Volterra systems of interacting species with symmetric community matrices. Cohen and Grossberg [1983] greatly extended this result to the following system

$$\dot{x}_i = a_i(x_i)\left[b_i(x_i) - \sum_{j=1}^{n} w_{ij} f_j(x_j) \right], \quad 1 \le i \le n, \quad (4.1.4)$$

subject to the following conditions:

(H1) symmetry: the matrix $[w_{ij}]$ is symmetric;

(H2) continuity: $a_i(\xi)$ is continuous in $\xi \ge 0$, $b_i(\xi)$ is continuous in $\xi > 0$;

(H3) positivity: $a_i(\xi) > 0$ if $\xi > 0$;

(H4) monotonicity: $f_i(\xi)$ is continuously differentiable and $f_i'(\xi) \ge 0$ for $\xi \ge 0$.

---

[1]The following is a standard result in integration theory (see, for example, McShane [1947]): Let $\phi : [a, b] \to \mathbb{R}$ be continuous and $\liminf_{h \to 0+} \frac{1}{h}[\phi(t+h) - \phi(t)] \le 0$ for all $t \in [a, b)$ except at most a countable number of points. Then ϕ is nonincreasing on $[a, b]$ and $\phi(t) - \phi(a) \le \int_a^b \phi'(s)ds$ for $t \in [a, b)$.

The above Cohen–Grossberg system can be used to represent a neural network where x_i is the activation of the ith neuron, $f_j(x_j)$ is the usual signal function and $W = (w_{ij})$ is the interconnection matrix of neurons. It means the activity of the ith neuron decreases if and only if the net input to the neuron exceeds a certain intrinsic function b_i of its activity, and the amplification factor $a_i(x_i)$ is positive.

An important example of (4.1.4) is the shunting on-center off-surround or cooperative-competitive feedback network. Recall that an on-center off-surround feedback neural network which obeys membrane equations is modeled by

$$\dot{x}_i = -A_i x_i + (B_i - x_i)[I_i + h_i(x_i)] - (x_i + C_i)\Big[J_i + \sum_{j=1, j\neq i}^{n} F_{ij} g_j(x_j)\Big], \quad (4.1.5)$$

where $1 \leq i \leq n$, x_i is the short-term memory activity of the ith cell (population), $-A_i x_i$ describes the passive decay of activity at rate A_i, $(B_i - x_i)[I_i + h_i(x_i)]$ describes how an excitatory input I_i and an excitatory feedback signal $h_i(x_i)$ increase the activity x_i, $-(x_i + C_i)\big[J_i + \sum_{j=1, j\neq i}^{n} F_{ij} g_j(x_j)\big]$ describes how an inhibitory input J_i and inhibitory feedback signal $F_{ij} g_j(x_j)$ from the jth cell decrease the activity x_i. We assume that $C_i > 0$ and $D_i > 0$ to describe the shunting or multiplicative effect. Note that each x_i can fluctuate within the finite interval $[-C_i, B_i]$, and the interval $[-C_1, B_1] \times \cdots \times [-C_n, B_n]$ is invariant for the flow defined by (4.1.5) (see the footnote in Section 3.4).

To write (4.1.5) in the form of Cohen–Grossberg system, we introduce the variables

$$y_i = x_i + C_i. \quad (4.1.6)$$

In terms of these variables, (4.1.5) can be rewritten as

$$\dot{y}_i = a_i(y_i)\Big[b_i(y_i) - \sum_{j=1, j\neq i}^{n} w_{ij} f_j(y_j)\Big] \quad (4.1.7)$$

with

$$\begin{cases} a_i(y_i) = y_i, \\ b_i(y_i) = \frac{1}{y_i}[A_i C_i - (A_i + J_i)y_i + (B_i + C_i - y_i)[I_i + h_i(y_i - C_i)]], \\ w_{ij} = F_{ij}, \\ f_i(y_i) = g_i(y_i - C_i). \end{cases}$$

Then (H2) and (H3) are satisfied. (H1) is satisfied if $F_{ij} = F_{ji}$ for $1 \leq i, j \leq n$. The monotonicity condition (H4) holds if $g_i' \geq 0$.

Recall that we require $x_i \geq -C_i$. This is equivalent to $y_i \geq 0$. It is therefore biologically reasonable to restrict our discussion to only nonnegative solutions of (4.1.7) or its more general form (4.1.4). Also, to apply LaSalle's Invariance Principle to (4.1.4) with $G = \text{int } \mathbb{R}^n_+$ (the open set of all vectors in $\mathbb{R}^n$ whose components are positive), we need to ensure all solutions of (4.1.4) starting in G remains in G and bounded for $t \geq 0$. The following conditions will be sufficient.

(H5) $w_{ij} \geq 0$, $f_i(\xi) \geq 0$ for $\xi \in \mathbb{R}$;

(H6) $\lim \sup_{\xi \to \infty}[b_i(\xi) - w_{ii} f_i(\xi)] < 0$;

(H7) either $\lim_{\xi \to 0^+} b_i(\xi) = \infty$ or

$$\lim_{\xi \to 0^+} b_i(\xi) < \infty \quad \text{and} \quad \int_0^\varepsilon \frac{d\xi}{a_i(\xi)} = \infty \quad \text{for some } \varepsilon > 0.$$

For system (4.1.7), (H5) holds provides $F_{ij} \geq 0$. As the signal functions h_i and g_i are bounded, $\lim \sup_{\xi \to \infty}[b_i(\xi) - w_{ii} f_i(\xi)] = -(A_i + J_i) - I_i < 0$. As $A_i C_i > 0$, we also have $\lim_{\xi \to 0^+} b_i(\xi) = \infty$.

The key to the Cohen–Grossberg's convergence theorem is the successful construction of a global Liapunov function for (4.1.4) given by

$$V(x) = -\sum_{i=1}^n \int_0^{x_i} b_i(\xi_i) f_i'(\xi_i) \, d\xi_i + \frac{1}{2} \sum_{j,k=1}^n w_{jk} f_j(x_j) f_k(x_k).$$

This is C^1-smooth, and by (4.1.3) we have

$$\begin{aligned}
\dot{V}_{(4.1.4)}(x) = &-\sum_{i=1}^n b_i(x_i) f_i'(x_i) a_i(x_i) \left[b_i(x_i) - \sum_{j=1}^n w_{ij} f_j(x_j)\right] \\
&+ \frac{1}{2} \sum_{j,k=1}^n w_{jk} f_j'(x_j) f_k(x_k) a_j(x_j) \left[b_j(x_j) - \sum_{l=1}^n w_{jl} f_l(x_l)\right] \\
&+ \frac{1}{2} \sum_{j,k=1}^n w_{jk} f_j(x_j) f_k'(x_k) a_k(x_k) \left[b_k(x_k) - \sum_{l=1}^n w_{kl} f_l(x_l)\right].
\end{aligned}$$

Relabeling the subindices, we get

$$\begin{aligned}
\dot{V}_{(4.1.4)}(x) = &-\sum_{i=1}^n b_i(x_i) f_i'(x_i) a_i(x_i) \left[b_i(x_i) - \sum_{j=1}^n w_{ij} f_j(x_j)\right] \\
&+ \frac{1}{2} \sum_{i=1}^n \left[\sum_{k=1}^n w_{ik} f_i'(x_i) f_k(x_k)\right] a_i(x_i) \left[b_i(x_i) - \sum_{j=1}^n w_{ij} f_j(x_j)\right] \\
&+ \frac{1}{2} \sum_{i=1}^n \left[\sum_{k=1}^n w_{ki} f_k(x_k) f_i'(x_i)\right] a_i(x_i) \left[b_i(x_i) - \sum_{j=1}^n w_{ij} f_j(x_j)\right].
\end{aligned}$$

Using the symmetry $w_{ik} = w_{ki}$, $1 \leq k$, $i \leq n$, we obtain

$$\dot{V}_{(4.1.4)}(x) = -\sum_{i=1}^n \left[b_i(x_i) - \sum_{j=1}^n w_{ij} f_j(x_j)\right] \left[b_i(x_i) - \sum_{j=1}^n w_{ij} f_j(x_j)\right] a_i(x_i) f_i'(x_i).$$

That is,

$$\dot{V}_{(4.1.4)}(x) = -\sum_{i=1}^{n} a_i(x_i) f_i'(x_i) \left[b_i(x_i) - \sum_{j=1}^{n} w_{ij} f_j(x_j) \right]^2. \qquad (4.1.8)$$

Consequently, $\dot{V}_{(4.1.4)}(x) \leq 0$ and V is indeed a Liapunov function of (4.1.4) on G. Note that system (4.1.4) can, in fact, be written in the gradient form

$$\dot{x} = A(x) \operatorname{grad} B(x) \qquad (4.1.9)$$

with

$$B(x) = -V(x) \qquad (4.1.10)$$

and

$$A_{ij}(x) = \frac{a_i(x_i)\delta_{ij}}{f_i'(x_i)}. \qquad (4.1.11)$$

To state a general convergence theorem, we define

$$\mathbb{R}_+^n = \{x \in \mathbb{R}^n; \ x_i \geq 0 \text{ for } 1 \leq i \leq n\},$$
$$\operatorname{int} \mathbb{R}_+^n = \{x \in \mathbb{R}^n; \ x_i > 0 \text{ for } 1 \leq i \leq n\}.$$

We say a solution $x : [0, \infty) \to \mathbb{R}^n$ of (4.1.4) is *admissible* if it is bounded and $x(t) \in \operatorname{int} \mathbb{R}_+^n$ for all $t \geq 0$. We will show that a solution of (4.1.4) subject to (H1)–(H7) is admissible if its initial value is in $\operatorname{int} \mathbb{R}_+^n$. We can now state the Cohen–Grossberg Convergence Theorem.

Theorem 4.1.2. *Assume that (H1)–(H7) are satisfied. Then for every solution $x :$ $[0, \infty) \to \mathbb{R}^n$ of (4.1.4) with $x(0) \in \operatorname{int} \mathbb{R}_+^n$, we have $x(t) \in \operatorname{int} \mathbb{R}^n$ for all $t \geq 0$ and $\omega(x(0)) \cap \operatorname{int} \mathbb{R}_+^n \subset M$, where M is the largest invariant set contained in the set*

$$E = \{y \in \operatorname{int} \mathbb{R}_+^n; \ \dot{V}_{(4.1.4)}(y) = 0\},$$

and

$$\dot{V}_{(4.1.4)}(y) = -\sum_{i=1}^{n} a_i(y_i) f_i'(y_i) \left[b_i(y_i) - \sum_{j=1}^{n} w_{ij} f_j(y_j) \right]^2.$$

In particular, if $f_i'(y_i) > 0$ for all $y_i > 0$ and $1 \leq i \leq n$, then the set E consists of only equilibria of (4.1.4) in $\operatorname{int} \mathbb{R}_+^n$.

Proof. By the LaSalle Invariance Principle, we only need to show that a solution $x : [0, \infty) \to \mathbb{R}^n$ of (4.1.4) is admissible if $x(0) \in \operatorname{int} \mathbb{R}_+^n$.

By (H6), there exists $L > 0$ such that

$$b_i(\xi) - w_{ii} f_i(\xi) < 0 \quad \text{for } 1 \leq i \leq n \text{ and } \xi \geq L. \qquad (4.1.12)$$

We claim that if $x(0) \in \text{int } \mathbb{R}_+^n$ then

$$0 < x_i(t) < \max\{x_i(0) + 1, L\} \quad \text{for } 1 \le i \le n \text{ and } t \ge 0.$$

If not, then there exist $1 \le i \le n$ and $T > 0$ such that either

$$x_i(T) = 0, \ 0 < x_i(t) \le \max\{x_i(0) + 1, L\} \text{ for } 0 < t < T \quad \text{and}$$
$$0 \le x_j(T) \le \max\{x_j(0) + 1, L\} \text{ for } 1 \le j \le n$$

or

$$x_i(T) = \max\{x_i(0) + 1, L\}, \ \dot{x}_i(T) \ge 0,$$
$$0 \le x_j(t) \le \max\{x_j(0) + 1, L\} \text{ for } 1 \le j \le n, 0 \le t \le T.$$

In the second case, we have from (H1)–(H5) and (4.1.12) that

$$\dot{x}_i(T) = a_i(x_i(T))\Big[b_i(x_i(T)) - \sum_{j=1}^n w_{ij} f_j(x_j(T))\Big]$$
$$\le a_i(x_i(T))[b_i(x_i(T)) - w_{ii} f_i(x_i(T))]$$
$$< 0,$$

a contradiction.

Consider the first case. If $\lim_{\xi \to 0^+} b_i(\xi) = \infty$ then as $t \to T^+$ we have

$$\dot{x}_i(t) = a_i(x_i(t))\Big[b_i(x_i(t)) - \sum_{j=1}^n w_{ij} f_j(x_j(t))\Big] > 0$$

since $x_i(t) \to 0^+$ and the terms $w_{ij} f_j(x_j(t))$ remain bounded, a contradiction. If $\lim_{\xi \to 0^+} b_i(\xi) < \infty$ and $\int_0^\varepsilon \frac{1}{a_i(\xi)} d\xi = \infty$ for some $\varepsilon > 0$, then

$$-\infty = \int_{x_i(0)}^0 \frac{d\xi}{a_i(\xi)} = \int_0^T \Big[b_i(x_i(t)) - \sum_{j=1}^n w_{ij} f_j(x_j(t))\Big] dt$$
$$> -\infty,$$

a contradiction.

It is simple to show that if (H1)–(H6) are satisfied and if $\lim_{\xi \to 0^+} b_i(\xi) = \infty$, then $x(0) \in \text{int } \mathbb{R}_+^n$ implies that $\liminf_{t \to \infty} x_i(t) > 0$ for every $1 \le i \le n$, and hence $\omega(x(0)) \subseteq \text{int } \mathbb{R}_+^n$.

Remark 4.1.3. Note that in the case where each function $f_i(x_i)$ is not strictly increasing, further argument is required in order to show that each admissible solution

approaches the set of equilibria. In fact, in a typical neural network, the inhibitory
feedback signal function $f_i(x_i)$ possesses an *inhibitory signal threshold* Γ_i such that

$$
\begin{cases}
f_i(x_i) = 0 & \text{if } x_i \leq \Gamma_i, \\
f_i'(x_i) > 0 & \text{if } x_i > \Gamma_i.
\end{cases}
$$

So each $f_i(x_i)$ is nondecreasing although not strictly increasing, function V continues
to be a Liapunov function and hence every admissible solution of (4.1.4) still converges
to the invariant set M. However, further analysis is required to guarantee that M
consists of equilibria only. The strategy developed in Cohen and Grossberg [1983] is
to decompose the variable x_i into suprathreshold and subthreshold variables, and then
to show how sets of suprathreshold equilibria can be used to characterize the ω-limit
set of each admissible solution of the full system (4.1.4), here we say the function $x_i(t)$
is *suprathreshold* at t if $x_i(t) > \Gamma_i$ and *subthreshold* at t if $x_i(t) \leq \Gamma_i$. At any time
t, suprathreshold variables receive signals only from other suprathreshold variables.
Because only suprathreshold variables signal other suprathreshold variables, one can
first restrict attention to all possible subsets of suprathreshold values that occur in
the ω-limit set of each admissible solution. Using the fact that each function f_i is
strictly increasing in the suprathreshold range, Cohen and Grossberg proved that the
suprathreshold subset corresponding to each ω-limit set defines an equilibrium point
of the subsystem of (4.1.4) that is constructed by eliminating all the subthreshold
variables of the ω-limit point. Under a weak additional assumption, they then showed
that the set of all such subsystem suprathreshold equilibrium points is countable, the
set of equilibrium points is totally disconnected and the ω-limit set of each admissible
solution is an equilibrium point. We refer to Cohen and Grossberg [1983] for more
details.

Remark 4.1.4. The network (4.1.5) is part of a mathematical classification theory
which characterizes how prescribed changes in system parameters alter the transformation from input patterns $(I_1, \ldots, I_n, J_1, \ldots, J_n)$ into activity pattern $(x_1, \ldots, x_n)$.
In addition to the study of prescribed transformations, the mathematical classification theory seeks the most general classes of networks where important general
processing requirements are guaranteed. Here, we study a class of networks which
transform arbitrary input patterns into activity patterns that are then stored in short-
term memory until a future perturbation resets the stored pattern. This property, also
called *global pattern formation*, means that given physically admissible input pattern
$(I_1, \ldots, I_n, J_1, \ldots, J_n)$ and initial activity pattern $(x_1(0), \ldots, x_n(0))$, the limit of
$(x_1(t), \ldots, x_n(t))$ as $t \to \infty$ exists.

Remark 4.1.5. System (4.1.4) includes the Volterra–Lotka equation

$$
\dot{x}_i = G_i x_i \left(1 - \sum_{j=1}^{n} H_{ij} x_j \right) \tag{4.1.13}
$$

and the Gilpin and Ayala system

$$\dot{x}_i = G_i x_i \left[1 - \left(\frac{x_i}{K_i} \right)^\theta - \sum_{j=1}^n H_{ij} \left(\frac{x_j}{K_j} \right) \right] \qquad (4.1.14)$$

from population biology; the dynamical model

$$\dot{x}_i(t) = -\frac{1}{2} x_i(t) - \sum_{j=1}^n \left[\frac{1}{\tau} x_i - r_{ij} \right]^+ k_{ij} + e_i \qquad (4.1.15)$$

of the Hartline–Ratliff–Miller equation

$$r_i(t) = e_i(t) - \sum_{j=1}^n \left[k_{ij} \frac{1}{\tau} \int_0^t e^{-(t-s)/\tau} r_j(s) \, ds - r_{ij} \right]^+$$

through the change of variable

$$x_i(t) = \int_0^t e^{-(t-s)} r_i(s) \, ds,$$

which describes the output r_i of the Limulus retina; the Eigen and Schuster equation

$$\dot{x}_i = x_i \left(m_i x_i^{p-1} - q \sum_{k=1}^n m_k x_k^v \right) \qquad (4.1.16)$$

from the theory of macromolecular evolution; the transformed dynamical analogue

$$\dot{y}_i = -y_i + \sum_{j=1}^n B_{ij} S(y_j) \qquad (4.1.17)$$

of the Brain-State-in-a Box (BSB) model introduced by Anderson, Silverstein, Ritz and Jones [1977]

$$\dot{x}_i = -x_i + S \left[\sum_{j=1}^n B_{ij} x_j \right]$$

under the transformation

$$y_i = \sum_{j=1}^n B_{ij} x_j;$$

as well as the dynamic McCulloch–Pitts model

$$\dot{y}_i = -y_i + \sum_{j=1}^n A_{ij} M(y_j), \qquad (4.1.18)$$

where

$$y_i = \sum_{j=1}^{n} A_{ij} x_j.$$

Note that the STM equation of the Boltzman machine (Ackley, Hinton and Sejnowitz [1985]) has the same form as that of the McCulloch–Pitts model with a logistic sigmoid function, while in BSB, S is a symmetric ramp function. Of course, the general additive STM equations and in particular, the Hopfield network can be written as a special form of (4.1.4) and we will discuss this type of networks in more detail in the next section.

Remark 4.1.6. In some applications of the general Cohen–Grossberg model (4.1.4), the coefficient matrix (w_{ij}) may be asymmetric, thereby rendering the general convergence theorem inapplicable. Asymmetric coefficients typically occur in problems relating to the learning and recognition of temporal order in behaviors. On the other hand, certain network models may have asymmetric interaction coefficients, yet reducible to the form (4.1.4) with symmetric interaction coefficients through a suitable change of variables. A typical example is the masking field model introduced by Grossberg to explain data about speech learning, word recognization, and the learning of adaptive sensory-motor plans. See Grossberg [1988] for details. See also next section for further discussions about convergence in a nonsymmetric network.

4.2   Convergence in networks with a nonsymmetric interconnection matrix

Assuming the LTM traces stabilize at constant values and lumping together the positive and negative feedback terms in the additive STM and passive delay LTM equation (2.1.10), we can rewrite the additive STM equation as

$$\frac{d}{dt} x_i = -A_i x_i + \sum_{j=1}^{n} z_{ij} f_j(x_j) + I_i, \quad 1 \le i \le n. \tag{4.2.1}$$

Substituting into (4.1.4) shows that

$$\begin{cases} a_i(x_i) = 1 \quad \text{(constant)}, \\ b_i(x) = I_i - A_i x \quad \text{(linear)}, \\ w_{ij} = -z_{ij}. \end{cases}$$

It is easy to verify that (H1)–(H4) are satisfied if $z_{ij} = z_{ji}$ for $1 \le i, j \le n$ and if $f_j' > 0$. Note also that if all f_j are bounded functions, then every solution of (4.2.1) is bounded. Then one can employ the same argument, using the Liapunov function[2]

$$V(x) = -\sum_{i=1}^{n} \int_0^{x_i} [I_i - A_i \xi_i] f_i'(\xi_i) d\xi_i - \sum_{j,k=1}^{n} z_{jk} f_j(x_j) f_k(x_k)$$

---

[2]In the Hopfield's electrical circuit interpretation (see Section 4.3), this is exactly the energy.

and applying LaSalle's Invariance Principle, to show that every solution of (4.2.1) is convergent to the set of equilibria.

The convergence theorem, however, requires the interconnection matrix $W = [w_{ij}]$ be symmetric. This requirement is justified when the synaptic strengths z_{ij} between cell v_i and cell v_j depend only on the intercellular distance. However, examples exist wherein the matrix W may be chosen as close to a symmetric matrix as one pleases, yet almost all trajectories persistently oscillate. A famous example is the May and Leonard's model of the voting paradox (see May and Leonard [1975]) described by the three-dimensional system

$$\begin{cases} \dot{x}_1 = x_1(1 - x_1 - \alpha x_2 - \beta x_3), \\ \dot{x}_2 = x_2(1 - \beta x_1 - x_2 - \alpha x_3), \\ \dot{x}_3 = x_3(1 - \alpha x_1 - \beta x_2 - x_3). \end{cases}$$

Grossberg [1978b] and Schuster, Sigmund and Wolff [1979] proved that if $\beta > 1 > \alpha$ and $\alpha + \beta > 2$, then all positive trajectories except the synchronized trajectories $x_1(0) = x_2(0) = x_3(0)$ persistently oscillate as $t \to \infty$, where the interconnection matrix

$$\begin{bmatrix} 1 & \alpha & \beta \\ \beta & 1 & \alpha \\ \alpha & \beta & 1 \end{bmatrix}$$

can be chosen arbitrarily close to a symmetric matrix by letting α and β approach 1.

Considerable effort has been made to remove or to weaken the symmetry condition in the Cohen–Grossberg convergence theorem. In this section, we are going to introduce a result of Gedeon [1999] and Fiedler and Gedeon [1998] in this direction.

We continue our discussion about the Cohen–Grossberg model

$$\dot{x}_i = a_i(x)\left[b_i(x_i) - \sum_{j=1}^{n} w_{ij} f_j(x_j)\right] \qquad (4.2.2)$$

without self-feedback (that is, $w_{ii} = 0$). We assume that $\frac{\partial f_j}{\partial x_j} > 0$ for all j, and that the functions $a_i(x)$ are nonnegative and $a_i(x) \geq 0$ for all $x \in \mathbb{R}^n$ and all i.

Our goal here is to drop the symmetry assumption on the structure matrix $W = [w_{ij}]$, by using results from combinatorial matrix theory.

To every $n \times n$ matrix W, we associate a *digraph* $D(W)$ containing n vertices and directed edges $\{\vec{e}_{ij}\}$, where there exists a directed $\vec{e}_{ij}$ with $i \neq j$ from vertex i to vertex j if and only if $w_{ij} \neq 0$. A *path* between vertex i_1 and vertex i_m is a collection of edges $\{\vec{e}_{i_1 i_2}, \vec{e}_{i_2 i_3}, \dots, \vec{e}_{i_{m-1} i_m}\}$. The above path is a cycle if $m \geq 3$, $i_1 = i_m$ and the vertices associated with $i_1, i_2, \dots, i_{m-1}$ are distinct from each other.

We say that the matrix W is *combinatorially symmetric*, if $w_{ij} \neq 0$ implies $w_{ji} \neq 0$. The matrix W is *sign symmetric*, if $w_{ij} \neq 0$ implies $w_{ij} w_{ji} > 0$. Observe that if W is combinatorially symmetric, $D(W)$ gives rise to an undirected graph $G(W)$ by identifying the edges $\vec{e}_{ij}$ and $\vec{e}_{ji}$ (and we will denote this identified edge by e_{ij}).

W is said to be *irreducible* if $G(W)$ is connected (Recall that a graph is *connected* if there always exists a path between two distinct vertices). In this section, we will assume that the matrix W in (4.2.1) is irreducible, The case where W is reducible will be dealt with in Section 4.5.

For a combinatorially symmetric matrix W, a slightly less restrictive assumption than sign symmetry condition is the following:

(A1)   There is a spanning tree T of $G(W)$ such that if $e_{ij} \in T$ then $w_{ij}w_{ji} > 0$.

Recall that a *tree* is a connected graph without cycles, and a *spanning tree* T of $G(W)$ is a tree containing every vertex of the graph $G(W)$.

For the given spanning tree T of $G(W)$, a *chord cycle* C_T is a collection of edges $e_{i_1 i_2}, e_{i_2 i_3}, \ldots, e_{i_k i_1}$ such that all edges, except one (called the *chord*) are in T. Note, that every cycle C in $G(W)$ corresponds to two cycles $C^1 = \vec{e}_{i_1 i_2}, \vec{e}_{i_2 i_3}, \ldots, \vec{e}_{i_k i_1}$ and $C^2 = \vec{e}_{i_2 i_1}, \vec{e}_{i_3 i_2}, \ldots, \vec{e}_{i_1 i_k}$ in the digraph $D(W)$. To every chord cycle $C_T \in G(W)$, we can associate two numbers:

$$p(C_T) := \Pi_{C_T^1} w_{ij} \quad \text{and} \quad \bar{p}(C_T) := \Pi_{C_T^2} w_{ji}.$$

We assume that

(A2) $p(C_T) = \bar{p}(C_T)$ for all chord cycles C_T.

The following result is due to Gedeon [1999]. See also Fiedler and Gedeon [1998] for a slightly weaker result.

Theorem 4.2.1. *Consider system* (4.2.2) *with assumptions* (A1) *and* (A2). *Then there exists a Liapunov function* $V : \mathbb{R}^n \to \mathbb{R}$, *non-increasing along the solutions and strictly decreasing along all non-equilibrium solutions of* (4.2.2) *(consequently, every bounded solution of* (4.2.2) *is convergent to the set of equilibria).*

In the above result, we replaced the symmetry assumption by (A1) and (A2). Observe that in the voting paradox model the condition (A2) is not satisfied: along the cycle $C := e_{12}, e_{23}, e_{31}$, we have

$$p(C) = \alpha^3 \neq \beta^3 = \bar{p}(C).$$

Observe that the assumption (A2) is always satisfied in the absence of cycles, *i.e.*, when the graph $G(W)$ is a tree.

Clearly, if W is sign symmetric and if $\Pi_C w_{ij} = \Pi_C w_{ji}$ along every cycle C (a condition required in Fiedler and Gedeon [1998]), then (A1) and (A2) are satisfied. we should point out, however, that in (A2) it is required to check certain conditions only for chord cycles of the graph. In a graph with a lot of edges, this makes a difference than the requirement that $\Pi_C w_{ij} = \Pi_C w_{ji}$ along every cycle C. For example, if the graph of interactions is a full graph with n vertices, then one needs to check $(n^2 - 3n + 2)/2$ chord cycles to verify assumption (A2), while there is $O(n!)$ cycles in such graph.

To illustrate this advantage, we consider a special case of system (4.2.1), where we assume that $w_{ij} w_{ji} > 0$ for $|i - j| = 1$. This implies that the system satisfies (A1) with a tree generated by the edges e_{ij} with $|i - j| = 1$. Assume further that $w_{kl} = w_{lk} = 0$ for some $l > k + 1$, and the rest of the system satisfies assumption (A2). Then it is always possible to choose weights $w_{kl} \neq 0$ and $w_{lk} \neq 0$ so that (A2) is satisfied. To see that, observe that we only need to check whether the weights along the chord cycle $e_{kl} e_{l,l-1} e_{l-1,l-2} \cdots e_{k+1,k}$ satisfy (A2). It follows that it is sufficient to choose w_{kl} and w_{lk} so that

$$\frac{w_{kl}}{w_{lk}} = \frac{w_{k,k+1} w_{k+1,k+2} \cdots w_{l-1,l}}{w_{k+1,k} w_{k+2,k+1} \cdots w_{l,l-1}}.$$

In particular, if system (4.2.1) is *tridiagonal* in the sense that $w_{ij} = 0$ for $|i - j| > 1$, and if this system satisfies (A1), we can add an arbitrary number of connections with appropriate weights so that the resulting system satisfies the assumptions of Theorem 4.2.1.

Proof of Theorem 4.2.1.   We first prove an abstract combinatorial result: *Let W be a real, combinatorially symmetric, irreducible matrix. Then there is a positive diagonal matrix D such that DW is a symmetric matrix if* (A1) *and* (A2) *are satisfied.*

To justify the claim, we seek to define numbers v_i, $i = 1, \ldots n$, such that

$$\frac{w_{ij}}{w_{ji}} = \exp(v_i - v_j) \quad \text{for all edges } e_{ij} \in G(W). \tag{4.2.3}$$

To construct the collection $\{v_i\}_{i=1}^n$ we pick v_1 and compute v_{i_1} for all i_1 with the property that $e_{1i_1} \in T$, using (4.2.3). We then use (4.2.3) to compute v_{i_2} for all i_2 with the property that $e_{i_1 i_2} \in T$ for some i_1 considered above. Repeating this process and due to the irreducibility of W, we can construct all v_i. Obviously, (4.2.3) is satisfied along all the edges of T. It remains to consider an edge $e_{kl} \in (G \setminus T)$. Since T is a spanning tree of $G(W)$, there is a path P in T from vertex k to vertex l. Thus $C = P \cup e_{kl}$ forms a chord cycle and by (A2), along this cycle we have $p(C) = \bar{p}(C)$, which we write as

$$\Pi_C \frac{w_{ij}}{w_{ji}} = 1.$$

Since (4.2.3) holds along edges in P this implies

$$1 = \Pi_C w_{ij} / \Pi_C w_{ji} = \frac{w_{kl}}{w_{lk}} \Pi_P \exp(v_i - v_j).$$

Simplifying the right-hand side, we get

$$\frac{w_{kl}}{w_{lk}} = \exp(v_k - v_l).$$

Since e_{kl} was arbitrary, collection $\{v_i\}$ satisfies (4.2.3) for all edges in $G(W)$.

Let $d_i := \exp(-v_i)$ and let $D = \text{diag}(d_i)$. Let $Q = DW$. Observe that

$$q_{ij}/q_{ji} = \frac{d_i}{d_j} \frac{w_{ij}}{w_{ji}} = \exp(-v_i)\exp(v_j)\exp(v_i - v_j) = 1$$

and so Q is a symmetric matrix. This proves the claim.

We now construct the Liapunov function. We use the symmetric matrix $Q = [q_{ij}]$ and the diagonal matrix $D = \text{diag}(d_i)$ from the claim to define the function $V : \mathbb{R}^n \to \mathbb{R}$ by

$$V(x) := -\sum_{i=1}^{n}\left(\gamma_i(x_i) - \sum_{j=1}^{n} q_{ij} f_i(x_i) f_j(x_j)\right),$$

where $\gamma_i = 2d_i \int^{x_i}\left(\frac{df_i}{d\zeta_i} b_i\right)(\zeta_i)\, d\zeta_i$. Abbreviating $f_i' = \frac{df_i}{dx_i}$ to simplify notation, we have

$$\dot{V}_{(4.2.2)}(x) = -\sum_{i=1}^{n}\left(\frac{d\gamma_i}{dx_i}\dot{x}_i - \sum_{j=1}^{n} q_{ij}(f_i' f_j(x_j)\dot{x}_i + f_j' f_i(x_i)\dot{x}_j)\right)$$

$$= -\sum_{i=1}^{n}\left(2d_i f_i' b_i \dot{x}_i - \sum_{j=1}^{n}(q_{ij} f_i' f_j(x_j)\dot{x}_i + q_{ji} f_i' f_j(x_j)\dot{x}_i)\right),$$

here we have rearranged the summations

$$\sum_{i=1}^{n}\sum_{j=1}^{n}[q_{ij} f_i' f_j(x_j)\dot{x}_i + q_{ij} f_j' f_i(x_i)\dot{x}_j] = \sum_{i=1}^{n}\sum_{j=1}^{n}[q_{ij} f_i' f_j(x_j)\dot{x}_i + q_{ji} f_i' f_j(x_j)\dot{x}_i].$$

Therefore,

$$\dot{V}_{(4.2.2)}(x) = -\sum_{i=1}^{n}\left(2d_i f_i' \dot{x}_i\left(b_i - \sum_{j=1}^{n}\frac{q_{ij} + q_{ji}}{2d_i} f_j(x_j)\right)\right)$$

$$= -\sum_{i=1}^{n}\left(2d_i f_i'(x_i)\dot{x}_i\left(b_i - \sum_{j=1}^{n}\frac{2q_{ij}}{2d_i} f_j(x_j)\right)\right)$$

$$= -\sum_{i=1}^{n} 2d_i f_i' a_i(x)\left[b_i - \sum_{j=1}^{n} w_{ij} f_j(x_j)\right]^2,$$

where we used equation (4.2.2) and that $w_{ij} = q_{ij}/d_i$. To finish the computation, we observe that d_i are positive constants, $f_i' > 0$ and $a_i(x) \geq 0$. Thus $\dot{V} \leq 0$. Since $2d_i f_i'$ is strictly positive, $\dot{V}_{(4.2.2)}(x) = 0$ if and only if $a_i(x)[b_i - \sum_{j=1}^{n} w_{ij} f_j(x_j)]^2 = 0$ for every i. This is equivalent to

$$a_i(x)\left(b_i - \sum_{j=1}^{n} w_{ij} f_j(x_j)\right) = 0$$

for every i and, consequently, $\dot{x}_i = 0$ for all i. Hence, $\dot{V}_{(4.2.2)}(x) = 0$ if and only if x is an equilibrium. This completes the proof of Theorem 4.2.1.

4.3 Implementation of CAM: Hopfield networks

We now follow the central idea in the Cohen–Grossberg convergence theorem to construct a network of n neurons $v_1, \ldots, v_n$ with fixed synaptic coupling coefficients to perform associative memory for the numerically coded memories $\{Z^1, \ldots, Z^M\}$, where

$$Z^j = (Z^j_1, \ldots, Z^j_n)^T$$

and Z^j_i are given numbers, $1 \leq i \leq n$, $1 \leq j \leq M$.

In general, the STM equation takes the form

$$\frac{dx}{dt} = f(x, Z^1, \ldots, Z^M), \tag{4.3.1}$$

where $x = (x_1, \ldots, x_n)^T$ and $f(\cdot, Z^1, \ldots, Z^M) : \mathbb{R}^n \to \mathbb{R}^n$ is a function to be determined. The discussions on CAM in last two sections suggest that we seek for a Liapunov (energy) function $P(x, Z^1, \ldots, Z^M)$ so that

$$\frac{\partial P}{\partial x_i} = -f_i(x, Z^1, \ldots, Z^M) \tag{4.3.2}$$

and the vectors $(Z^1, \ldots, Z^M)$ are the extreme of P as a function of $x \in \mathbb{R}^n$.

The simplest choice of such an energy function is a quadratic function. For example, if we have a single memory $(M = 1)$, then it is natural to let

$$P(x, Z^1) = (a/2)x^T x - (Z^1, x) \quad \text{for some } a > 0,$$

giving

$$\frac{dx}{dt} = -ax + Z^1. \tag{4.3.3}$$

This system relaxes to $x = a^{-1}Z^1$ for every initial condition $x(0)$.

In the case where we have two coded memories, we consider the system as follows

$$\frac{dx}{dt} = -ax + C_1(x, Z^1)Z^1 + C_2(x, Z^2)Z^2, \tag{4.3.4}$$

where C_1 and C_2 are coefficients to be determined. The basic principle for the choice of C_i is to choose C_i to measure the closeness of x to Z^i so that if $a = 1$ (equivalent to rescaling the variables) and if $x(0)$ is close to Z^i, then the system relaxes to Z^i.

A simple closeness measure is using the inner product of $\operatorname{sgn} x$ and Z^1. Namely,

$$C_1(x, Z^1) = (\operatorname{sgn} x, Z^1) = \sum_{i=1}^{n} (\operatorname{sgn} x_i) Z^1_i. \tag{4.3.5}$$

Note that when $x = Z^1$, one obtains the Manhattan norm

$$C_1(Z^1, Z^1) = (\text{sgn } Z^1, Z^1) = \sum_{i=1}^{n} |Z_i^1| = \|Z^1\|_1. \tag{4.3.6}$$

Similarly, one defines $C_2(x, Z^2) = (\text{sgn } x, Z^2)$. Assume the two memories Z^1 and Z^2 are distinguishable in the sense that if x is close to Z^i then $|C_i(x, Z^i)| \gg |C_j(x, Z^j)|$ for $i \neq j$. Then near Z^i, (4.3.4) relaxes towards Z^i.

Generalizing to M memories, the system becomes

$$\frac{dx}{dt} = -ax + Z^1(\text{sgn } x, Z^1) + \cdots + Z^M(\text{sgn } x, Z^M). \tag{4.3.7}$$

Note that

$$Z^j(\text{sgn } x, Z^j) = Z^j[(Z^j)^T \text{ sgn } x] = Z^j(Z^j)^T \text{ sgn } x.$$

Therefore, factoring out $\text{sgn } x$ in (4.3.7) gives

$$\frac{dx}{dt} = -ax + W \text{ sgn } x, \tag{4.3.8}$$

where

$$W = \sum_{j=1}^{M} Z^j(Z^j)^T. \tag{4.3.9}$$

The corresponding energy P satisfies

$$\left(\frac{\partial P}{\partial x_1}, \ldots, \frac{\partial P}{\partial x_n} \right)^T = ax - W \text{ sgn } x. \tag{4.3.10}$$

Therefore, P is the sum of an upward-opening quadratic term $\frac{1}{2}ax^T x$ and an downward perturbation P' with $\left(\frac{\partial P'}{\partial x_1}, \ldots, \frac{\partial P'}{\partial x_n} \right)^T = -W \text{ sgn } x$.

Recall that we hope that each Z^i is a stable equilibrium of (4.3.7). It is then desirable to have $aZ^j \approx Z^j(\text{sgn } Z^j, Z^j) = \|Z^j\|_1 Z^j$ from which it follows that $a \approx \|Z^j\|_1$ for $1 \leq j \leq M$. In other words, the memory vectors should be sufficiently distinguishable by the closeness measurable and should be distributed on the surface of an n-dimensional sphere. This is called *simplex signal set* in digital communication theory.

Other forms may be used which lead to different function coefficients. For example, the weighted norm

$$C_1(x, z) = (\text{sgn } x, z)_V = z^T V \text{ sgn } x,$$

where V is a positive definite matrix, allowing weighting the significance of each element.

We emphasize that the true set of equilibria of (4.3.7) is not exactly the set of memories $\{Z^1, \ldots, Z^M\}$, and how close these two sets are depends on the choice of the set of $\{Z^1, \ldots, Z^M\}$ and the choice of the inner product.

To build a machine with the additive STM model, we write (4.3.8) as

$$\dot{x}_i = -ax_i + \sum_{j=1}^{n} w_{ij} \operatorname{sgn} x_j. \tag{4.3.11}$$

Replacing the $\operatorname{sgn}(x_j)$ function with a general signal function $f_j(x_j)$, we get

$$\dot{x}_i = -ax_i + \sum_{j=1}^{n} w_{ij} f_j(x_j), \tag{4.3.12}$$

which is exactly the additive model (4.2.1) discussed in last section.

We conclude this section with a short note about the hardware implementation of (4.3.12) once $\{w_{ij}\}$ is determined. See Müller and Reinhardt [1991] and Hopfield [1982, 1984] for more detailed discussions.

It is relatively easy to implement (4.3.12) in an artificial neural network consisting of electronic neurons (amplifiers) interconnected through a matrix of resistors. Here an electronic neuron, the building block of the network, consists of a nonlinear amplifier which transforms an input signal u_i into the output signal v_i, and the input impedance of the amplifier unit is described by the combination of a resistor ρ_i and a capacitor C_i. See Figure 4.3.1 for the circuit diagram of a single neuron.

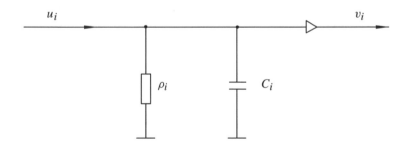

Figure 4.3.1. The circuit diagram of a single neuron.

We assume the input-output relation is completely characterized by a voltage amplification function

$$v_i = f_i(u_i).$$

The synaptic connections of the network are represented by resistors R_{ij} which connect the output terminal of the amplifier j with the input part of the neuron i. See Figure 4.3.2 for an electronic circuit diagram of a network of 3 neurons. In order that the network can function properly the resistances R_{ij} must be able to take on negative values. This can be realized through a slight modification of Figure 4.3.2, namely, the amplifiers are supplied with an inverting output line which produces the signal $-v_j$. The number of rows in the resistor matrix is doubled, and whenever a negative value of R_{ij} is needed this is realized by using an ordinary resistor which is connected to the inverting output line. See Figure 4.3.3 for a circuit with inverting amplifiers.

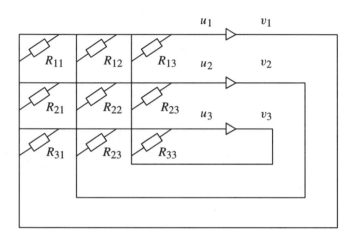

Figure 4.3.2. The electronic circuit diagram for a network of 3 neurons.

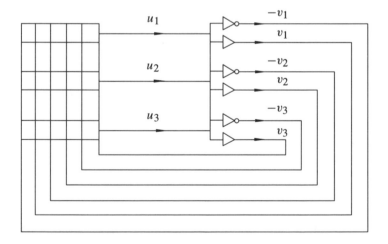

Figure 4.3.3. The circuit diagram for a network of 3 neurons with inverting amplifiers, denoted by ▷∘.

The time evolution of the signals of the network is described by the Kirchhoff's law. Namely, the strength of the incoming and outgoing current at the amplifier input port must balance.[3] Consequently, we arrive at

$$C_i \frac{du_i}{dt} + \frac{u_i}{\rho_i} = \sum_{j=1}^{n} \frac{1}{R_{ij}} (v_j - u_i).$$

Let

$$\frac{1}{R_i} = \frac{1}{\rho_i} + \sum_{j=1}^{n} \frac{1}{R_{ij}}.$$

We get

$$C_i R_i \frac{du_i}{dt} + u_i = \sum_{j=1}^{n} \frac{R_i}{R_{ij}} v_j$$

or

$$T_i \frac{du_i}{dt} + u_i = \sum_{j=1}^{n} w_{ij} f_j(u_j),$$

here $T_i = C_i R_i$ and $w_{ij} = \frac{R_i}{R_{ij}}$ denote the local relaxation time and the synaptic strength.

4.4 Generic convergence in monotone networks

In this section, we describe a class of network whose interconnection matrix satisfies certain "sign symmetry" and "irreducibility" properties, for which the convergence is guaranteed for almost every trajectory.

We will need some general results for systems of ordinary differential equations in $\mathbb{R}^n$ which have the property that the flow preserves a partial ordering "$\leq_K$" in $\mathbb{R}^n$ generated by one of the orthants K of $\mathbb{R}^n$, and further references can be found in Hirsch [1987, 1989]. The general material presented here is taken from Smith [1988]. We consider an autonomous system of ordinary differential equations

$$\dot{x} = g(x), \quad x \in U \subseteq \mathbb{R}^n, \tag{4.4.1}$$

here g is a continuously differentiable function defined in a suitable open convex set U of $\mathbb{R}^n$. We write $\Phi(t, x_0)$ for the solution of (4.4.1) which satisfies $\Phi(0, x_0) = x_0$. Let $K = \{x \in \mathbb{R}^n; (-1)^{m_i} x_i \geq 0, 1 \leq i \leq n\}$ be an orthant of $\mathbb{R}^n$, where $m_i \in \{0, 1\}$. Then K is a cone in $\mathbb{R}^n$ and generates a partial ordering "$\leq_K$" in $\mathbb{R}^n$ in the usual fashion: $x \leq_K y$ if and only if $y - x \in K$. We say that the flow of (4.4.1) preserves $\leq_K$ in case that whenever x and y lie in U and $x \leq_K y$, then $\Phi(t, x) \leq_K \Phi(t, y)$ for all $t \geq 0$ for which both solutions are defined.

---

[3]The book of Hoppensteadt [1997] contains a brief discussion about electrical circuits

The well-known result, often attributed to Kamke but in fact due to Müller, gives sufficient conditions for the flow of (4.4.1) to preserve the usual partial ordering on $\mathbb{R}^n$ induced by the cone $\mathbb{R}^n_+ = \{x \in \mathbb{R}^n; x_i \geq 0 \text{ for } 1 \leq i \leq n\}$. These conditions require, roughly speaking, that g_i is nonincreasing in $x_j \in U$ for $j \neq i$. More precisely, we have the following sufficient conditions:

Lemma 4.4.1. *If* $Dg(x) = \left(\frac{\partial g_i(x)}{\partial x_j}\right)$ *has nonnegative off-diagonal elements for every* $x \in U$*, then the flow of* (4.4.1) *preserves the ordering* $\overset{\leq}{=}_K$ *with* $K = \mathbb{R}^n_+$.

We refer to Kamke [1932] or Müller [1926] for the proof.

The above result was extended, by Burton and Whyburn [1952], to the general case where $K_m = \{x \in \mathbb{R}^n; (-1)^{m_i} x_i \geq 0, 1 \leq i \leq n\}$ with $m = (m_1, \ldots, m_n)$ and $m_i \in \{0, 1\}$ for $1 \leq i \leq n$.

Lemma 4.4.2. *The flow of* (4.4.1) *preserves the partial ordering* $\overset{\leq}{=}_{K_m}$ *if* $P_m Dg(x) P_m$ *has nonnegative off-diagonal elements for every* $x \in U$*, where* $P_m = \text{diag}((-1)^{m_1}, \ldots, (-1)^{m_n})$.

In what follows, we call (4.4.1) a *type K_m monotone system* if its flow preserves the partial ordering $\overset{\leq}{=}_{K_m}$, and we will drop the subscript m if no confusion arises.

Remark 4.4.3. The idea in the proof of Lemma 4.4.2 is simple: namely, we note that if (4.4.1) is type K monotone, then the change of variables $y = Px$ in (4.4.1) yields a system $\dot{y} = h(y)$, $h \in C^1(PU)$, $h(y) = Pg(Py)$, for which the flow preserves the usual partial ordering $\overset{\leq}{=}$ on $\mathbb{R}^n$ and $Dg(y)$ has nonnegative off-diagonal elements in PU. So, in theory, one only needs to consider system (4.4.1) where $Dg(x)$ has nonnegative off-diagonal elements for every $x \in U$. In the applications, however, one often has some intuition concerning the system (4.4.1) which may be obscured by making the above change of variables.

A "canonical" form for type K monotone systems can be obtained by permuting equations and variables in (4.4.1) in the same way so that we can arrive at a system (4.4.1) where $x = (x_1, x_2) \in \mathbb{R}^k \times \mathbb{R}^{n-k}$, $g = (g_1, g_2)$ and

$$Dg(x) = \begin{bmatrix} \frac{\partial g_1}{\partial x_1} & \frac{\partial g_1}{\partial x_2} \\ \frac{\partial g_2}{\partial x_1} & \frac{\partial g_2}{\partial x_2} \end{bmatrix}$$

with $\partial g_1/\partial x_1$ and $\partial g_2/\partial x_2$ having nonnegative off-diagonal elements and $\partial g_1/\partial x_2 \leq 0$ and $\partial g_2/\partial x_1 \leq 0$. In these coordinates, the flow of (4.4.1) preserves the ordering induced by the cone $\mathbb{R}^k_+ \times (-\mathbb{R}^{n-k}_+)$.

Lemma 4.4.2 leads to a useful algorithm for testing a system (4.4.1) in order to see if it is type K monotone for some m (and for finding m). Although for small values of n, it is usually a trivial matter to identify a type K monotone system, when n is large it can be a nontrivial matter. To describe a simple algorithm for detecting type K monotonicity, let us first reinterpret the sufficient condition of Lemma 4.4.2 as

$$(-1)^{m_i+m_j} \frac{\partial g_i}{\partial x_j}(x) \geq 0, \quad i \neq j, \ x \in U \tag{4.4.2}$$

and noting that it requires "sign stability" of the off-diagonal elements of $Dg(x)$, i.e., $\partial 9_i/\partial x_j$ cannot have opposite signs in U for each i, j, $i \neq j$, and it requires "sign symmetry" of $Dg(x)$ in the sense that $(\partial g_i/\partial x_j)(\partial g_j/\partial x_i) \geq 0$ in U. These conditions are not sufficient, however, as will be seen later.

We now develop a simple algorithm to determine when a particular system is type K monotone and which partial ordering is preserved by the corresponding flow. From Lemma 4.4.2, it is clear that we must compute $Dg(x)$. First, we check the Jacobian for sign stability of the off-diagonal elements, i.e., one must establish that for given $i \neq j$ either $\partial g_i/\partial x_j(x) \leq 0$ for all $x \in U$ or that $\partial g_i/\partial x_j(x) \geq 0$ for all $x \in U$. Having done this, and assuming that (4.4.1) passes this test, we must next consider the requirement of sign symmetry of Dg in U. In other words, $\partial g_i/\partial x_j(x)\partial g_j/\partial x_i(y) \geq 0$ should hold for all $i \neq j$, x, $y \in U$. This can usually be checked simply by a look at Dg. Again, we assume that (4.4.1) has passed this test of sign symmetry. Now it suffices to consider only the entries of Dg above the main diagonal.

For each $i < j$, $(i, j) \in \mathbb{N} \times \mathbb{N}$, define $s_{ij} \in \mathbb{Z}_2 = (\{0, 1\} + (\mod 2))$ by the following

$$s_{ij} = \begin{cases} 0 & \text{if } \frac{\partial g_i}{\partial x_j} > 0 \text{ for some } x \in U, \\ 1 & \text{if } \frac{\partial g_i}{\partial x_j} < 0 \text{ for some } x \in U, \\ \text{arbitrary} & \text{if } g_i \text{ is independent of } x_j. \end{cases}$$

Since we are assuming Dg is sign stable, s_{ij} is well defined. Now consider the solvability of the system of $n(n-1)/2$ equations in the n unknowns $m_i \in \mathbb{Z}_2, 1 \leq i \leq n$

$$\begin{aligned}
m_1 + m_2 &= s_{12}, \\
m_1 + m_3 &= s_{13}, \\
&\vdots \\
m_1 + m_n &= s_{1n}, \\
m_2 + m_3 &= s_{23}, \quad (\mod 2). \\
&\vdots \\
m_2 + m_n &= s_{2n}, \\
&\vdots \\
m_{n-1} + m_n &= s_{n-1,n}.
\end{aligned} \tag{4.4.3}$$

If there is a solution $m \in \mathbb{Z}_2^n$ (for some choice of the s_{ij} which were arbitrarily assigned), then (4.4.1) is type $\tilde{K}_m$ monotone for this value of m. Equivalently, since an equation involving an arbitrarily assigned s_{ij} can always be satisfied by an appropriate choice of s_{ij} depending on m_i and m_j, we can delete such equation from (4.4.3) and consider the solvability of the remaining equations only. Note that if (4.4.3) is solved by $m \in \mathbb{Z}_2^n$ then $m + \underline{1}$ also solves (4.4.3) where $\underline{1} = (1, 1, \ldots, 1)$ since the kernel of the left-hand side of (4.4.3) is just $\{\underline{0}, \underline{1}\}$. This is not at all surprising since m and $m + \underline{1}$ correspond to orthants K_m and $-K_m$ respectively and a flow is easily seen to preserve one partial ordering if and only if it preserves the other.

There is a graph theoretic test for type K monotonicity. We assume $Dg(x)$ is sign stable in U. Consider the graph Γ with vertices 1, 2, 3, $\ldots$, n. If $\partial g_i / \partial x_j$ is not identically zero in U, put a directed edge $\overrightarrow{e}_{ij}$ from j to i (arrow pointing to i) and attach the sign of $\partial g_i / \partial x_j$ to the edge. Then (4.4.1) is type K monotone if and only if for every cycle in Γ the number of minus signs is even.

The asymptotic behavior of a typical solution of a type K monotone system is convergence to an equilibrium. The following gives a relatively simple sufficient condition for a bounded solution to converge to an equilibrium.

Lemma 4.4.4. *Let a positive orbit of a type K monotone system $O^+(x) = \{\Phi(t, x);$ $t \geq 0\}$ have compact closure in U and assume $g(x) \overset{\geq}{=}_K 0\,(g(x) \overset{\leq}{=}_K 0)$. Then $\Phi(t, x)$ tends to an equilibrium as $t \to \infty$.*

The details of the proof can be found in Selgrade [1980], we give here only a brief informal sketch. In the case with $g(x) \overset{\geq}{=}_K 0$, it is immediate from the monotonicity properties of g and that $x + K_m$ is positively invariant for (4.4.1) (it is easiest, and it suffices, to see this for $K_m = \mathbb{R}_+^n$). Hence $\Phi(t, x) \overset{\geq}{=}_{K_m} x$ for small $t \geq 0$ and thus $\Phi(t + s, x) \overset{\geq}{=}_{K_m} \Phi(s, x)$ whenever $s \geq 0$ by the monotonicity of the flow. It follows that the components of $\Phi(t, x)$ are monotone functions of $t \in \mathbb{R}$ and this yields the convergence for $\Phi(t, x)$ as $t \to \infty$.

To obtain convergence without requiring $g(x) \overset{\geq}{=}_K 0$, we need a slightly stronger monotonicity property. We say (4.4.1) is irreducible if $Dg(x)$ is an irreducible matrix for every $x \in U$. Recall that an $n \times n$ matrix is *irreducible* if it does *not* leave invariant any proper nontrivial subspace generated by a subset of the standard basis vectors for $\mathbb{R}^n$ or, equivalently, if one cannot put the matrix in the form

$$\begin{bmatrix} A & B \\ 0 & C \end{bmatrix},$$

where A and C are square matrices by a reordering of the standard basis. In other words, a matrix $W = (w_{ij})$ is irreducible if for every nonempty, proper subset I of the set $N = \{1, \ldots n\}$, there is $i \in I$ and $j \in N \setminus I$ such that $w_{ij} \neq 0$. Somewhat inaccurately, (4.4.1) is irreducible if it cannot be decomposed into two subsystems, one of which does not depend on the other.

The flow of an irreducible type K monotone system has a strong order-preserving property described below.

Lemma 4.4.5. *If (4.4.1) is an irreducible type K monotone system, then for two* *distinct points x and y of U with $x \leq_K y$ but $x \neq y$ we have $\Phi(t, x) <_K \Phi(t, y)$ for $t > 0$, where for $x, y \in \mathbb{R}^n$, $x <_K y$ if and only if $y - x$ is in the interior of K.*

The proof can be found in Hirsch [1984, 1988].

The recent development on monotone dynamical systems shows that almost every bounded trajectory of an irreducible type K monotone system is convergent to the set of equilibria. The following is due to Hirsch [1988].

Theorem 4.4.6. *Let Φ be the flow of an irreducible type K monotone system with the property that all positive orbits have compact closures. Assume that the set of equilibria of (4.4.1) is a discrete set in U. Then the set Y of points $x \in U$ for which $\Phi(t, x)$ does not converge to an equilibrium of (4.4.1) has Lebesgue measure zero.*

A much stronger assertion than that of Theorem 4.4.6 can be made in the case that more is assumed about the properties of the flow. Let E denote the set of equilibria of (4.4.1). If $e \in E$, let $B(e)$ denote the *basin of attraction* of e, namely, $B(e) = \{x \in U; \Phi(t, x) \to e$ as $t \to \infty\}$. The following result is a consequence of Theorem 9.4 of Hirsch [1988]:

Theorem 4.4.7. *Let X_0 be an open, bounded, positively invariant set for an irreducible type K monotone system whose closure (denoted by $\overline{X_0}$) contains only a finite number of equilibria. Then*

$$\bigcup_{e \in U \cap \overline{X_0}} \text{int}(B(e))$$

is open and dense in X_0, where $\text{int}(S)$ denotes the interior of S.

It is not difficult to see that $\text{int}(B(e)) = \emptyset$ if e is linearly unstable (there exists an eigenvalue of the Jacobian with positive real part).[4]

The above results of Hirsch suggest that the asymptotic behavior of a typical solution of a type K monotone irreducible system is the convergence to an equilibrium and the union of the basins of attraction of stable equilibria is dense in the phase space. Thus it will be particularly important to determine the stability type of each equilibrium of (4.4.1). If x_0 is an equilibrium of (4.4.1) then the matrix $A = Dg(x_0)$ is the coefficient matrix of the variational equation about x_0. By Lemma 4.4.2, A is similar to a matrix whose off-diagonal elements are nonnegative. We will call any matrix A, having the property that the off-diagonal elements of $P_m A P_m$ are nonnegative for some m, a type K matrix. We write $s(A)$ for the stability modulus of an $n \times n$ matrix,

---

[4]Assume x^* is an equilibrium of (4.4.1), that is, $g(x^*) = 0$. We say that x^* is stable if for every $\varepsilon > 0$ there exists $\delta = \delta(\varepsilon) > 0$ such that $|x_0 - x^*| < \delta$ implies $|\Phi(t, x_0) - x^*| < \varepsilon$ for all $t \geq 0$. If x^* is not stable, then we say that x^* is unstable. A stable equilibrium x^* is said to be asymptotically stable if there exists $\delta_0 > 0$ such that $|x_0 - x^*| < \delta_0$ implies that $\Phi(t, x_0) \to x^*$ as $t \to \infty$. The linearization principle gives the stability information of x^* from the eigenvalues of the Jacobian $A = Dg(x^*)$. Namely, it says that x^* is asymptotically stable if all eigenvalues of A have negative real parts, and x^* is unstable if A has an eigenvalue with positive real part. See Hale [1969].

namely, $s(A) = \max \operatorname{Re} \lambda$ where λ runs over the spectrum of A. We call A a stable matrix if $s(A) < 0$. The following result summarizes some of the special properties of type K matrices.

Theorem 4.4.8. *Let A be a type K matrix. Then $s(A)$ is an eigenvalue of A and there exists a corresponding eigenvector $v \in K$. The following are equivalent:*

(i) $s(A) < 0$.

(ii) *There exists $u >_K 0$ such that $Au <_K 0$.*

(iii) $-A^{-1} \geqq_K 0$.

(iv)

$$(-1)^k \begin{vmatrix} a_{11} & |a_{12}| & \cdots & & \cdots & |a_{1k}| \\ |a_{21}| & a_{22} & |a_{23}| & & \cdots & |a_{2k}| \\ \vdots & & & & & \vdots \\ |a_{k1}| & \cdots & & \cdots & |a_{k,k-1}| & a_{kk} \end{vmatrix} > 0, \quad 1 \le k \le n.$$

(v) *There exists $d > 0$ such that the symmetric matrix $\operatorname{diag}(d) A + A^T \operatorname{diag}(d)$ is negative definite, where $\operatorname{diag}(d)$ is the diagonal matrix with $d_1, \ldots, d_n$ down the main diagonal.*

We remark that the inequality $B \geqq_K 0$ for a matrix means precisely that B maps the orthant K into itself.

The Perron–Frobenius theorem about the spectrum of a nonnegative matrix lies behind many of the assertions of Theorem 4.4.8. For a proof of most of Theorem 4.4.8 and for many other interesting facts about type K matrices, we refer the reader to Berman and Plemmons [1979, Chap. 6].

In order to interpret the results of Theorem 4.4.8, it is useful (and causes no loss of generality) to assume that A has nonnegative off-diagonal elements, in which case the partial ordering is the usual one. Notice that by (ii), A can be stable only if each diagonal element is negative and dominates the other entries in that row. This result is certainly intuitive. It is quite remarkable that (iv) is precisely the test for negative definiteness of a symmetric matrix (ignoring the absolute values in the general case). We have a straightforward test for stability of a type K monotone system, in contrast to the case of a general matrix.

We now identify certain classes of additive models of neural networks whose solution operators preserve a partial ordering $\leqq_{K_m}$. Consider

$$\dot{x}_i = -\alpha_i x_i + \sum_{j=1}^{n} w_{ij} f_j(x_j), \tag{4.4.4}$$

where we assume $\alpha_i > 0$ and the signal function $f_j : \mathbb{R} \to \mathbb{R}$ is C^1-smooth and $\frac{\partial f_j(x_j)}{\partial x_j} > 0$ for all $1 \leq j \leq n$ and $x_j \in \mathbb{R}$. Let

$$g_i(x) = -\alpha_i x_i + \sum_{j=1}^{n} w_{ij} f_j(x_j).$$

Then

$$\frac{\partial g_i(x)}{\partial x_j} = w_{ij} f'_j(x_j)$$

and thus $\frac{\partial g_i(x)}{\partial x_j}$ and w_{ij} have the same signs. Clearly, system (4.4.4) is always sign stable. It is sign symmetric if and only if

$$w_{ij} w_{ji} \geq 0 \qquad 1 \leq i, j \leq n. \tag{4.4.5}$$

We will always assume that (4.4.5) is satisfied in the following discussion.

Example 4.4.9 ($n = 2$). In the case where $n = 2$, system (4.4.4) is sign symmetric if and only if $w_{12} w_{21} \geq 0$. If, in addition, the network is fully connected (irreducible), then $w_{12} w_{21} > 0$. This is true if either the output of one neuron excites another (w_{12}, $w_{21} > 0$) or the output of one neuron inhibits another (w_{12}, $w_{21} < 0$). If $w_{12} > 0$ and $w_{21} > 0$, then (4.4.4) is type K monotone with $K = \mathbb{R}_+^2$; and if $w_{12} < 0$ and $w_{21} < 0$, then (4.4.4) is type K monotone with $K = \{(x_1, x_2)^T ; x_1 \geq 0, x_2 \leq 0\}$. In either case, it is well known that all bounded solutions tend to equilibria as $t \to \infty$. See, for example, Hirsch and Smith [1974].

Example 4.4.10 ($n = 3$). Network (4.4.4) is type K_m monotone if and only if the equations

$$\begin{aligned} m_1 + m_2 &= s_{12} \\ m_1 + m_3 &= s_{13} \qquad \text{(mod 2)} \\ m_2 + m_3 &= s_{23} \end{aligned}$$

are solvable in $\mathbb{Z}_2^3$, where

$$(-1)^{s_{ij}} = \text{sign}\,[w_{ij} f'_j(x_j)] = \text{sign}\,w_{ij}.$$

These equations can be solved for $m \in \mathbb{Z}_2^3$ if and only if $s_{12} + s_{13} + s_{23} = 0 \,(\text{mod } 2)$. Hence (4.4.4) with $n = 3$ and the condition (4.4.5) is type K_m monotone if and only if

$$\text{sign}\,w_{12}\text{sign}\,w_{23}\text{sign}\,w_{31} \geq 0,$$

or equivalently

$$w_{12} w_{23} w_{31} \geq 0.$$

Example 4.4.11 (Nearest neighbor interactions on a circle or line segment). This is the case where $w_{ij} = 0$ if $|i - j| > 1$ for $i, j = 1, 2, \ldots, n \pmod{n}$. This describes the interaction of neurons arranged on a circle where each neuron interacts only with its nearest neighbors.

Then $Dg(x)$ is tridiagonal except for the entries $\partial g_1/\partial x_n = w_{1n} f_n'(x_n)$ in the upper right-hand corner and $\partial g_n/\partial x_1 = w_{n1} f_1'(x_1)$ in the lower left-hand corner. Assuming (4.4.5), we will establish conditions for (4.4.4) to be type K_m monotone for some m. Equations (4.4.3) reduce to the following

$$
\begin{aligned}
m_1 + m_2 &= s_{12}, \\
m_1 + m_n &= s_{1n}, \\
m_2 + m_3 &= s_{23}, \\
&\;\;\vdots \\
m_{n-1} + m_n &= s_{n-1,n}.
\end{aligned}
\tag{4.4.6}
$$

Ignoring the second equation in (4.4.6) we can solve the remaining equations for m_2, $\ldots, m_n$ in terms of m_1: $m_j = s_{12} + s_{23} + \cdots + s_{j-1,j} + m_1 \pmod{2}$, $2 \le j \le n$. The second equation will be satisfied if and only if

$$
s_{1n} + s_{12} + s_{23} + \cdots + s_{n-1,n} = 0 \quad \pmod{2}.
$$

This is the solvability condition. If it is satisfied then (4.4.4) is type K_m monotone for m as calculated above. The solvability condition can be restated in terms of the derivatives as follows:

$$
\prod_{i=1}^{n} w_{i,i+1} f_{i+1}'(x_{i+1}) \ge 0.
\tag{4.4.7}
$$

Using the same discussion as for $n = 3$, we know (4.4.7) is equivalent to

$$
w_{12} w_{23} \ldots w_{n-1,n} w_{n,1} \ge 0.
\tag{4.4.8}
$$

Observe that if $w_{i,i+1} = 0$ in U for some $i \in \{0, 1, \ldots, n\}$, then (4.4.8) is automatically satisfied. In this case by renumbering we may as well assume $w_{1n} = w_{n1} = 0$ in U, or g_1 is independent of x_n and g_n is independent of x_1. We refer to such systems as *sign symmetric tridiagonal systems* since $Dg(x)$ is a tridiagonal matrix. As we have seen, sign-symmetric tridiagonal systems are always type K_m monotone systems. Smillie [1984] has obtained the following result for such systems.

Theorem 4.4.12. *Suppose each f_i is $n - 1$ times differentiable[5] and (4.4.4) is a sign-symmetric and irreducible system with $w_{1n} = w_{n1} = 0$. If $x(t)$ is a bounded solution for $t \ge 0$ then $\lim_{t \to \infty} x(t)$ exists and is an equilibrium point.*

---

[5] It is still an open question whether this additional smoothness condition is indeed needed.

We now consider the general case of (4.4.4). Note that since $f_j'(x_j) > 0$ for $j = 1, 2, \ldots, n$ and $x_j \in \mathbb{R}$, the sufficient condition (4.4.2) becomes

$$(-1)^{m_i + m_j} w_{ij} \geq 0. \tag{4.4.9}$$

Consequently, (4.4.4) is type K monotone if and only if the interconnection matrix W is a K matrix, which is equivalent to that (4.4.9) holds. In other words, if we regard neurons with excitatory interconnections as "friends" and neurons with inhibitory interconnections as "enemies", then (4.4.4) is type K monotone if "friends of friends are friends, friends of enemies are enemies and enemies of enemies are friends". Note also that (4.4.4) is irreducible if and only if W is irreducible. Consequently, we have

Theorem 4.4.13. *Assume that the interconnection matrix W is type K monotone and irreducible. Assume that the set of equilibria of (4.4.4) is discrete, and that all trajectories are bounded (this is the case if $\alpha_i > 0$ and if each signal function f_i is bounded for $1 \leq i \leq n$), then the set Y of points $x \in \mathbb{R}^n$ for which $\Phi(t, x)$ does not converge to an equilibrium of (4.4.4) has Lebesgue measure 0, and $\bigcup_{e \in E} \text{int}(B(e))$ is open and dense in $\mathbb{R}^n$, where E is the set of equilibria.*

Chapter 5

Signal transmission delays

As already noted in Chapters 1 and 2, time delays always occur in the signal transmission between neurons in biological neural networks. Such delays are omnipresent in artificial neural networks and, in particular, in the electronic hardware implementation, due to the finite propagation velocity of neural signals and due to the finite switching speed of amplifiers (neurons).

In this chapter, we will discuss the effect of signal delays on the network dynamics. An important issue to be addressed is how signal delays change the stability of equilibria, causing nonlinear oscillations and inducing periodic solutions. It will be shown that increasing the delay is among many mechanisms to create a network that exhibits periodic oscillations. While the existence of periodic solutions creates some problems in neural network applications such as CAM, periodic sequences of neural impulses are of fundamental importance for the control of motor body functions, such as the heartbeat, which occurs with great regularity almost three billion times during an average person's life. Irregularities in the cardiac rhythm are fortunately rare, but cause grave concern if they do occur. The questions how neural networks can sustain highly periodic activity for a long period of time and what makes them fail under certain conditions are thus of vital interest (Müller and Reinhart [1991]).

We will also address other problems related to the dynamic behaviors of networks with delay such as synchronization and phase-locked oscillations. The ultimate goal is to completely understand the dynamic behaviors of all trajectories of the models — systems of delay differential equations. As will be illustrated, information about these behaviors is coded by the so-called global attractor. We will give some relatively complete description of the topological and geometric structures of the attractor and dynamics of the restricted flow generated by the model equation on the attractor for some networks with simple connection topology.

We conclude this introductory section by listing several main types of phenomena observed in networks with delay, and we will provide detailed discussions later.

1. *Delay-independent stability in contractive neural networks.* Although we are interested in changes induced by the delay, we should mention that, as will be illustrated in Section 5.2, in globally asymptotically stable contractive networks, the delay does not alter the asymptotic behavior of the system.

2. *Almost quasiconvergence in cooperative irreducible networks.* This is another situation in which the delay has limited effect on the long-term behavior of the system. Recall that an irreducible network is one in which there is a directed

path connecting any pair of neurons. It is referred to as cooperative when, possibly after a change of variables, all interactions are excitatory. As will be discussed in Section 5.2, such systems, with or without delay, generate an eventually strongly monotone semiflow, so that the asymptotic behavior of the network with delay is essentially the same as that of the corresponding network without delay.

3. *Delay-induced periodic oscillations in frustrated networks.* It is known that the presence of long delays may render unstable an otherwise stable equilibrium point, for example through a Hopf bifurcation, thus leading to stable periodic oscillations. As will be noted in Sections 5.3 and 5.4, for neural networks an important necessary condition for the occurrence of such phenomenon is the presence of frustration, that is, a directed pathway from a neuron to itself such that the product of the connection weights along the path is negative. Therefore, simple models for frustrated networks are the single self-inhibiting neuron, and two-neuron (A, B) networks with excitation from neuron A to neuron B and inhibition from neuron B to neuron A.

4. *Delay-induced chaotic oscillations.* It should be emphasized that delay-induced oscillation need not be periodic, and even a two-neuron network can provide a tractable prototype for the study of delay-induced chaos in neural networks (Gilli [1993]).

5. *Delay-induced synchronization and desynchronization.* In a network of identical neurons and in the absence of synaptic delays, the asymptotic behaviors of the network can be characterized by the convergence to the set of synchronized equilibria. Delay, however, can cause stable phase-locked oscillations and other asynchronous behaviors. This will be illustrated in Section 5.9.

6. *Delay-induced change of the domain of attraction.* As will be shown in Section 5.7, even for a cooperative irreducible system whose quasiconvergence is guaranteed, delay may affect the shape and structure of the basins of attraction of equilibria which has significant impact on its application to CAM.

7. *Delay-induced transient oscillations.* All the phenomena described above are related to the influence of delay on the long-term behaviors of the system. It should be noted that the delay also affects the transient behavior of networks. As will be shown in Section 5.6, for some neural networks delay may lead to long-lasting transient oscillations whose duration increases exponentially with the delay.

5.1   Neural networks with delay and basic theory

Recall that in the derivation of the model equation (see Section 4.3)

$$C_i R_i \frac{du_i}{dt} = -u_i + \sum_{j=1}^{n} \frac{R_i}{R_{ij}} f_j(u_j)$$

for an artificial neural network, we implicitly assumed that the neurons communicate and respond instantaneously. Consideration of the finite switching speed of amplifiers requires that the input-output relation be replaced by

$$v_i = f_i(u_i(t - \tau_i))$$

with a positive constant $\tau_i > 0$, and thus we obtain the following system of delay differential equations

$$C_i R_i \frac{du_i(t)}{dt} = -u_i(t) + \sum_{j=1}^{n} \frac{R_i}{R_{ij}} f_j(u_j(t - \tau_j)), \qquad 1 \le i \le n.$$

In what follows, for the sake of simplicity, we assume that

$$C_i = C, \quad R_i = R, \quad 1 \le i \le n,$$

and thus all local relaxation times $C_i R_i = CR$ are the same. Rescaling the time delay with respect to the network's relaxation time and rescaling the synaptic connection by

$$x_i(t) = u_i(CRt), \quad r_j = \frac{\tau_j}{RC}, \quad w_{ij} = \frac{R}{R_{ij}},$$

we get

$$\frac{d}{dt} x_i(t) = -x_i(t) + \sum_{j=1}^{n} w_{ij} f_j(x_j(t - r_j)). \tag{5.1.1}$$

It is now easy to observe that it is the relative size of the delay r_j which determines the dynamics and the computational performance of the network, and designing a network to operate more quickly will increase this relative size of the delay.

In biological neural networks, if the axon is a thick myelinated one then the transportation of signals is fast. However, a considerable portion of the axons (estimated range from 20 to 40 percent) are very fine myelinated fibers or unmyelinated ones. These axons play quite an important role since they give rise to long delays, up to 100–200 ms (Herz, Sulzer, Kühn and van Hemmen [1989]). In addition, there may be (post)synaptic delays, so that long-term potential even occurs if the signal from the source neuron has arrived some time before the target neuron becomes active. In general, axonal signal propagation corresponds to a discrete time delay such as the one in (5.1.1), diffusion processes across the synapses and along the dendrites are modeled by delay distributions with single[1] peaks. So a biologically realistic model should take the following form

$$C_i \frac{\mathrm{d}}{\mathrm{d}t} u_i(t) = -\frac{1}{R_i} u_i(t) + \sum_{j=1}^{n} \int_0^\tau w_{ij}(s) f_j(u_j(t-s)) \, \mathrm{d}s + I_i(t), \qquad (5.1.2)$$

with $w_{ij} : [0, \tau] \to \mathbb{R}$ being given functions and I_i the external inputs. See Herz [1994], an der Heiden [1980] and Herz, Sulzer, Kühn and van Hemmen [1989].

We now describe some general qualitative and geometric results for delay differential equations, with emphasis on applications to model equation (5.1.1) or (5.1.2).

To specify a solution of the model equation (5.1.1), we need to prescribe the value of the solution on the initial interval $[-r, 0]$, with $r = \max_{1 \le j \le n} r_j$. It is therefore natural to introduce the following phase space

$$C = C([-r, 0]; \mathbb{R}^n) = \{\phi; \phi : [-r, 0] \to \mathbb{R}^n \text{ is continuous}\}.$$

This is a Banach space if it is equipped with the super-norm

$$\|\phi\| = \sup_{s \in [-r, 0]} |\phi(s)|, \quad \phi \in C,$$

where $|\cdot|$ denotes the usual Euclidean norm.[2]

The following notation is standard and crucial in an abstract formulation of a delay differential equation: Let $A > 0$ and $a \in \mathbb{R}$ be given constants and let $x : [a - r, a + A] \to \mathbb{R}^n$ be a given continuous mapping. Then, for each $t \in [a, a + A]$, we define $x_t : [-r, 0] \to \mathbb{R}^n$ by

$$x_t(s) = x(t + s), \quad s \in [-r, 0].$$

It can be easily verified that $x_t \in C$ and the mapping $[a, a + A] \ni t \mapsto x_t \in C$ is continuous. Geometrically, x_t is the segment of the graph of x restricted to $[t - r, t]$, switched back to the initial interval $[-r, 0]$. See Figure 5.1.1.

Let $F : C \to \mathbb{R}^n$ be a given continuous mapping, then the following relation

$$\dot{x}(t) = \frac{\mathrm{d}}{\mathrm{d}t} x(t) = F(x_t) \qquad (5.1.3)$$

is called a *delay differential equation*. To write the network model (5.1.1) in the form of (5.1.3), we simply let $F = (F_1, \ldots, F_n)^T$ and

$$F_i(\phi) = -\phi_i(0) + \sum_{j=1}^{n} w_{ij} f_j(\phi_j(-r_j)), \quad \phi = (\phi_1, \ldots, \phi_n)^T \in C. \qquad (5.1.4)$$

---

[1]Distributed time lags with multiple peaks may be used to include pathways via interneurons that are not explicitly represented in the model.

[2]Another natural choice of the phase space is the product

$$C([-r_1, 0]; \mathbb{R}) \times C([-r_2, 0]; \mathbb{R}) \times \cdots \times C([-r_n, 0]; \mathbb{R}).$$

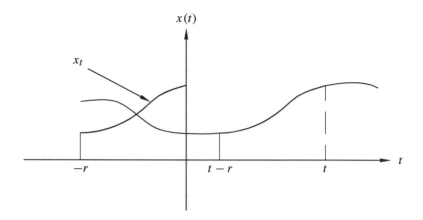

Figure 5.1.1. x_t is the segment of x on $[t-r, t]$, switched back to the initial interval.

Clearly, for the above chosen F and for some mapping $x : \mathbb{R} \to \mathbb{R}^n$ and $t \in \mathbb{R}$, we get

$$F_i(x_t) = -(x_i)_t(0) + \sum_{j=1}^{n} w_{ij} f_j((x_j)_t(-r_j))$$

$$= -x_i(t) + \sum_{j=1}^{n} w_{ij} f(x_j(t - r_j)).$$

Similarly, we can write (5.1.2) as a special form of (5.1.3). The initial value of the delay differential equation (5.1.3) is

$$x_{t_0} = \phi \tag{5.1.5}$$

for a given $t_0 \in \mathbb{R}$ (the *initial time*) and a given $\phi \in C$ (the *initial value*). A *solution* of the initial value problem (5.1.3) and (5.1.5) is a continuous mapping $x : [t_0 - r, t_0 + \beta) \to \mathbb{R}^n$ for some $\beta > 0$ such that $x_{t_0} = \phi$, the restriction of x to the interval $(t_0, t_0 + \beta)$ is continuously differentiable and satisfies (5.1.3). The interval $[t_0 - r, t_0 + \beta)$ with the maximal β is called the *maximal interval of existence*.

To state the basic theory for the delay differential equation (5.1.3), we introduce the concept of a Lipschitz mapping. We say that $F : C \to \mathbb{R}^n$ is *locally Lipschitz continuous* if for every $M > 0$ there exists $L_M > 0$ such that

$$|F(\phi) - F(\psi)| \leq L_M \|\phi - \psi\| \quad \text{if } \|\phi\| \leq M, \|\psi\| \leq M.$$

It is easy to verify that if each $f_j : \mathbb{R} \to \mathbb{R}$ in (5.1.1) is locally Lipschitz continuous, then so is $F : C \to \mathbb{R}^n$ defined in (5.1.4).

The following theorem gives the fundamental existence, uniqueness, continuous dependence and continuation of the initial value problem of (5.1.3).

Theorem 5.1.1. *Assume that $F : C \to \mathbb{R}^n$ is locally Lipschitz continuous. Then for each $\phi \in C$ and for each $t_0 \in \mathbb{R}$, the initial value problem (5.1.3) and (5.1.5) has one and only one solution on an interval $[t_0 - r, t_0 + \beta)$ for some $\beta > 0$. If, we further assume that F maps bounded sets of C into bounded sets of $\mathbb{R}^n$, then the maximal interval of existence $I(t_0, \phi) = [t_0 - r, t_0 + \beta)$ of the solution to (5.1.3) and (5.1.5) satisfies that either $\beta = \infty$ or $\lim_{t \to (t_0 + \beta)^-} \|x_t(t_0, \phi)\| = \infty$, here and in what follows, we use $x(t_0, \phi)$ to denote the unique solution of (5.1.3) and (5.1.5). Finally, $x_t(t_0, \phi)$ is continuous with respect to (t, t_0, ϕ) in the following sense: for every given $t_0 \in \mathbb{R}$ and $\phi \in C$, if $[t_0 - r, t_0 + \beta)$ is the maximal interval of existence of $x(t_0, \phi)$, then for each $\varepsilon > 0$ and $\beta^* \in (0, \beta)$ there exists $\delta > 0$ so that for every $t^* \in \mathbb{R}$ and $\phi^* \in C$ with $|t^* - t_0| < \delta$ and $\|\phi^* - \phi\| < \delta$, the solution $x(t^*, \phi^*)$ exists on $[t^* - r, t^* + \beta^*]$ and $\|x_{\tilde{t}}(t^*, \phi^*) - x_t(t_0, \phi)\| < \varepsilon$ for all $\tilde{t}, t \in [\max\{t_0, t^*\}, t^* + \beta^*]$ with $|\tilde{t} - t| < \delta$.*

We refer to Driver [1977] for an elementary introduction to delay differential equations and to Hale [1977], Hale and Verduyn Lunel [1993], and Diekmann, Verduyn Lunel, van Gils and Walther [1995] for the proof and detailed discussions related to the above theorem.

Assuming each $f_j : \mathbb{R} \to \mathbb{R}$ is continuously differentiable, then it is easy to verify that the mapping $F : C \to \mathbb{R}^n$ defined in (5.1.4) is locally Lipschitz continuous and maps bounded sets of C into bounded sets of $\mathbb{R}^n$. Therefore, an initial value problem for (5.1.1) has a unique solution which is continuous with respect to the initial time and the initial value.

According to Theorem 5.1.1, to ensure the solution of the initial value problem (5.1.3) subject to (5.1.5) is defined for all $t \geq t_0$, it is sufficient to show that the solution remains in a bounded set for t in any interval $[t_0, t_0 + \alpha]$ with $\alpha > 0$. In particular, for equation (5.1.1), if each signal function $f_j : \mathbb{R} \to \mathbb{R}$ is bounded, then by using the variation-of-constants formula, every solution x satisfies

$$x_i(t) = e^{-(t-t_0)} x_i(t_0) + \int_{t_0}^{t} e^{-(t-s)} \sum_{j=1}^{n} w_{ij} f_j(x_j(s - r_j)) \, ds \qquad (5.1.6)$$

and thus remain in a bounded set of $\mathbb{R}^n$ in its maximal interval of existence, and so the solution is defined for all $t \geq t_0$. In fact, formula (5.1.6) also implies the boundedness of the solution on the whole interval $[t_0, \infty)$.

In what follows, we will assume that $F : C \to \mathbb{R}^n$ is locally Lipschitz continuous, maps bounded sets of C into bounded sets of $\mathbb{R}^n$ and that every solution of (5.1.3) subject to $x_0 = \phi \in C$ is defined for all $t \geq 0$. Let $\Phi : [0, \infty) \times C \to C$ be defined by

$$\Phi(t, \phi) = x_t(0, \phi).$$

Then Φ satisfies the following properties:

(i) Φ is continuous;

(ii) $\Phi(0, \phi) = \phi$ for all $\phi \in C$;

(iii) $\Phi(t + s, \phi) = \Phi(t, \Phi(s, \phi))$ for all $t, s \geq 0$ and for all $\phi \in C$.

In other words, Φ is a *semiflow* on C. For a given $\phi \in C$, the mapping $\Phi(\cdot, \phi)$: $[0, \infty) \to C$ is called the *trajectory* through ϕ and the image $O^+(\phi) = \{\Phi(t, \phi); t \geq 0\}$ is called the *positive orbit* through ϕ. The ω-*limit set* of ϕ is defined as

$$\omega(\phi) = \{\psi \in C; \text{ there exists a sequence of nonnegative real numbers } \{t_n\}_{n=1}^{\infty}$$
$$\text{such that } t_n \to \infty \text{ and } \Phi(t_n, \phi) \to \psi \text{ as } n \to \infty\}.$$

It is known that if the orbit $O^+(\phi)$ is bounded then $\omega(\phi)$ is nonempty, compact, connected, invariant and $\text{dist}(\Phi(t, \phi), \omega(\phi)) \to 0$ as $t \to \infty$, where

$$\text{dist}(\psi, \omega(\phi)) = \inf_{\xi \in \omega(\phi)} \|\psi - \xi\|,$$

and a set $M \subseteq C$ is said to be invariant if for every $\xi \in M$ there exists a C^1-mapping $x : \mathbb{R} \to \mathbb{R}^n$ such that $x_0 = \xi$, $\dot{x}(t) = F(x_t)$ and $x_t \in M$ for all $t \in \mathbb{R}$. For such a mapping, we say $\{x_t; t \leq 0\}$ is a *negative orbit* through ϕ and we can define the α-limit set of such a negative orbit in a similar way to that for ω-limit set, except replacing $t_n \to \infty$ by $t_n \to -\infty$.

An *equilibrium* is a point $\varphi \in C$ such that $\Phi(t, \varphi) = \varphi$ for all $t \geq 0$. It can be easily shown that an equilibrium φ must be a constant mapping $\varphi(\theta) = x^*$ for some $x^* \in \mathbb{R}^n$ and $\theta \in [-r, 0]$. Denote by $\hat{x}^*$ the constant mapping in $[-r, 0]$ with the constant value $x^* \in \mathbb{R}^n$. Then $\hat{x}^*$ is an equilibrium if and only if $F(\hat{x}^*) = 0$. Clearly, $\omega(\hat{x}^*) = O^+(\hat{x}^*) = \{\hat{x}^*\}$. We say an equilibrium $\hat{x}^*$ is *stable* if for any $\varepsilon > 0$ there exists $\delta > 0$ such that $\|\varphi - \hat{x}^*\| < \delta$ implies $\|\Phi(t, \varphi) - \hat{x}^*\| < \varepsilon$ for all $t \geq 0$; *asymptotically stable* if it is stable and if there exists $\varepsilon_0 > 0$ such that $\|\varphi - \hat{x}^*\| < \varepsilon_0$ implies $\|\Phi(t, \varphi) - \hat{x}^*\| \to 0$ as $t \to \infty$; *unstable* if it is not stable. The stability of an equilibrium is usually determined by the distribution of zeros for a transcendental equation associated with the linearization of (5.1.1) at the equilibrium $\hat{x}^*$, as will be shown in subsequent sections. The *basin of attraction* of an equilibrium x^* is the set of initial value $\varphi \in C$ such that $\Phi(t, \varphi) \to x^*$ as $t \to \infty$. Note that in the remaining part of this chapter, we will not distinguish between x^* and $\hat{x}^*$.

If there exists $T > 0$ such that $\Phi(t + T, \varphi) = \varphi$ for all $t \geq 0$, then we say $O^+(\varphi)$ is a T-periodic orbit. Clearly, in this case, we have

$$O^+(\varphi) = \omega(\varphi) = \{\Phi(t, \varphi); t \in [0, T]\}.$$

We say the T-periodic orbit $O^+(\varphi)$ is *stable* if for any $\varepsilon > 0$ there exists $\delta > 0$ so that $\text{dist}(\psi, O^+(\varphi)) < \delta$ implies $\text{dist}(\Phi(t, \psi), O^+(\varphi)) < \varepsilon$ for all $t \geq 0$; *orbitally (asymptotically) stable* if it is stable and if there exists $\varepsilon_0 > 0$ so that $\text{dist}(\psi, O^+(\varphi)) < \varepsilon_0$ implies $\lim_{t \to \infty} \text{dist}(\Phi(t, \psi), O^+(\varphi)) = 0$; *unstable* if it is not stable. The stability of a periodic orbit is determined by the spectrum (Floquet multipliers) of the so-called monodromy operator of a linear variational equation of (5.1.1) at the orbit, as will be seen in later sections.

If there exists a constant $B > 0$ such that $\lim \sup_{t \to \infty} \|\Phi(t, \varphi)\| \leq B$ for every $\phi \in C$, then we say the semiflow Φ is *point dissipative*. Using the variation-of-constants formula (5.1.6), we can easily show that the semiflow Φ generated by (5.1.1) is point dissipative if each f_j is bounded. It is known that if Φ is point dissipative, then there exists the smallest compact invariant set Ω such that for any bounded set U of C, we have

$$\lim_{t \to \infty} \sup_{\varphi \in U} \mathrm{dist}(\Phi(t, \varphi), \Omega) = 0. \qquad (5.1.7)$$

See Hale [1977]. Such a compact invariant set Ω is called the *(global) attractor* of the semiflow Φ. Clearly, it contains all limiting behaviors of all trajectories starting from a given bounded subset of C, and thus it is of great importance in applications of neural networks to describe the structures of Ω.

5.2   Global stability analysis

It is of importance to determine whether signal transmission delay can destabilize a network whose convergence is guaranteed in the absence of delay. In this section, we are going to show that signal delay does not alter the asymptotic behavior of a globally asymptotically stable contractive network.

Recall that multiple equilibria usually coexist in a Hopfield network with symmetric connection matrix and they correspond to local extrema of a Liapunov function. It is desirable to have multiple stable equilibria in the application of neural networks to CAM, but in other applications to signal and image processing involving optimization problems, it is useful to have a unique equilibrium whose basin of attraction is the whole phase space. In what follows, we describe a result of Gopalsamy and He [1994] which gives some sufficient conditions to ensure the existence and global attractivity of a unique equilibrium in a delayed network of neurons.

Consider the Hopfield network of neurons

$$\dot{x}_i = -\mu_i x_i + \sum_{j=1}^{n} w_{ij} f_j(x_j(t - \tau_{ij})) + I_i, \quad 1 \leq i \leq n \qquad (5.2.1)$$

with constants $\mu_i > 0$, connection matrix $W = (w_{ij})$, signal functions $f_j : \mathbb{R} \to \mathbb{R}$, signal delays $\tau_{ij} \geq 0$ and external inputs I_i which are assumed to be temporally uniform over the duration of interest (and thus are constants). We assume each signal function $f_j : \mathbb{R} \to \mathbb{R}$ satisfies $f_j(0) = 0$ and is globally Lipschitz continuous so that[3]

$$|f_j(x) - f_j(y)| \leq L_j |x - y| \quad \text{for } x, y \in \mathbb{R}, 1 \leq j \leq n \qquad (5.2.2)$$

---

[3]Note that we do not require the monotonicity of each signal function. It was shown in Morita [1993] and Yoshizawa, Morita and Amari [1993] that non-monotone signal functions may increase the capacity of a network for the purpose of CAM.

for given constants $L_j > 0$. Note that if $f_j : \mathbb{R} \to \mathbb{R}$ is C^1-smooth, monotonically increasing and with maximal slope at 0, then we can choose $L_j = f_j'(0)$.

Theorem 5.2.1. *Assume*

$$\alpha = \max_{1 \le i \le n} \frac{L_i}{\mu_i} \sum_{j=1}^{n} |w_{ji}| < 1. \tag{5.2.3}$$

Then for each given input $I = (I_1, \ldots, I_n)^T \in \mathbb{R}^n$, the network has one and only one equilibrium $x^* = (x_1^*, \ldots, x_n^*)^T \in \mathbb{R}^n$ such that every solution of (5.2.1) satisfies $\lim_{t \to \infty} x_i(t) = x_i^*$, $1 \le i \le n$.

Proof. Note that $x^* = (x_1^*, \ldots, x_n^*)^T$ is an equilibrium if and only if $y^* = (\mu_1 x_1^*, \ldots, \mu_n x_n^*)^T$ is a fixed point of the map $F : \mathbb{R}^n \to \mathbb{R}^n$ given by

$$F_i(y) = \sum_{j=1}^{n} w_{ij} f_j \left(\frac{y_j}{\mu_j} \right) + I_i, \quad 1 \le i \le n. \tag{5.2.4}$$

Let $\|z\| = \sum_{i=1}^{n} |z_i|$ denote the norm of $z \in \mathbb{R}^n$, we have

$$\|F(y) - F(z)\| = \sum_{i=1}^{n} |F_i(y) - F_i(z)|$$

$$\le \sum_{i=1}^{n} \sum_{j=1}^{n} |w_{ij}| \left| f_j \left(\frac{y_j}{\mu_j} \right) - f_j \left(\frac{z_j}{\mu_j} \right) \right|$$

$$\le \sum_{i=1}^{n} \sum_{j=1}^{n} |w_{ij}| \frac{L_j}{\mu_j} |y_j - z_j|$$

$$= \sum_{i=1}^{n} \left[\frac{L_i}{\mu_i} \sum_{j=1}^{n} |w_{ji}| \right] |y_i - z_i|$$

$$\le \alpha \|y - z\|.$$

Hence, F is a contraction in $\mathbb{R}^n$ and thus has a unique fixed point $x^* = (x_1^*, \ldots, x_n^*)^T \in \mathbb{R}^n$.

To establish the global attractivity, we let

$$y_i(t) = x_i(t) - x_i^*$$

and get

$$\dot{y}_i(t) = -\mu_i y_i(t) + \sum_{j=1}^{n} w_{ij} [f_j(y_j(t - \tau_{ij}) + x_i^*) - f_j(x_i^*)]. \tag{5.2.5}$$

Using the Lipschitz continuity (5.2.2), we can estimate the upper right derivative $D^+|y_i(t)|$ as[4]

$$D^+|y_i(t)| \leq -\mu_i|y_i(t)| + \sum_{j=1}^{n}|w_{ij}|L_j|y_j(t - \tau_{ij})|.$$

Consider a Liapunov functional $V(t) = V(y)(t)$ defined by

$$V(y)(t) = \sum_{i=1}^{n}\left[|y_i(t)| + \sum_{j=1}^{n}|w_{ij}|L_j\int_{t-\tau_{ij}}^{t}|y_j(s)|ds\right].$$

Then

$$D^+V(t) \leq \sum_{i=1}^{n}\left[-\mu_i|y_i(t)| + \sum_{j=1}^{n}|w_{ij}|L_j|y_j(t - \tau_{ij})|\right]$$

$$+ \sum_{i=1}^{n}\sum_{j=1}^{n}|w_{ij}|L_j[|y_j(t)| - |y_j(t - \tau_{ij})|]$$

$$= -\sum_{i=1}^{n}\mu_i|y_i(t)| + \sum_{i=1}^{n}\sum_{j=1}^{n}|w_{ji}|L_i|y_i(t)|]$$

$$= \sum_{i=1}^{n}\left[-\mu_i + L_i\sum_{j=1}^{n}|w_{ji}|\right]|y_i(t)|$$

$$= -\sum_{i=1}^{n}\mu_i\left[1 - \frac{L_i}{\mu_i}\sum_{j=1}^{n}|w_{ji}|\right]|y_i(t)|$$

$$\leq -\min_{1\leq i\leq n}\mu_i\alpha\|y(t)\|.$$

Therefore, we have

$$V(y)(t) + \min_{1\leq i\leq n}\mu_i\alpha\int_{0}^{t}\|y(t)\|\,dt \leq V(y)(0),$$

from which it follows that $\|y\|$ is bounded on $(0, \infty)$ and $\|y\| \in L^1[0, \infty)$. By equation (5.2.5) and the Lipschitz continuity of f_j, we have the boundedness of $|y_i'(t)|$ on $(0, \infty)$. Hence, y_i is uniformly continuous on $(0, \infty)$. This, together with $\|y\| \in L^1(0, \infty)$, implies that $\|y(t)\| \to 0$ as $t \to \infty$. That is, $\lim_{t\to\infty}x_i(t) = x_i^*$ for $1 \leq i \leq n$. This completes the proof.

Using the following Liapunov functional

$$V(y)(t) = \sum_{i=1}^{n}\left[\frac{1}{2}y_i^2(t) + \frac{1}{2}\sum_{j=1}^{n}L_j|w_{ij}|\int_{t-\tau_{ij}}^{t}y_j^2(s)\,ds\right] \qquad (5.2.6)$$

---

[4]The upper right derivative $D^+z(t)$ of a given function $z : [a, b] \to \mathbb{R}$ at $t \in [a, b]$ is defined as $\limsup_{h\to 0+} h^{-1}[z(t + h) - z(t)]$. It is known in the integration theory that if $D^+z(t) \leq h(t)$ for a Lebesgue integrable function $h : [a, b] \to \mathbb{R}$ then $z(t) \leq z(a) + \int_a^t h(s)ds$ for all $t \in [a, b]$.

and a similar argument, Cao and Zhou [1998] derived another set of sufficient conditions for the uniqueness and global attractivity of the equilibrium of (5.2.1) described as follows:

$$\max_{1 \le i \le n} \frac{1}{\mu_i} \sum_{j=1}^{n} [L_j |w_{ij}| + L_i |w_{ji}|] < 2, \tag{5.2.7}$$

$$\max_{1 \le i \le n} \frac{1}{\mu_i} \sum_{j=1}^{n} [L_j^2 |w_{ij}| + L_i^2 |w_{ji}|] < 2. \tag{5.2.8}$$

Results of the above nature can also be found in van den Driessche and Zou [1998].

Despite the aforementioned progress, it is still quite remote to have satisfactory solutions to the following

Open Problems. Assume the network (5.2.1) has a unique equilibrium whose basin of attraction is the whole phase space if all delays τ_{ij} are equal to zero. Under which conditions, we can claim that (i) this equilibrium remains to be the global attractor for all nonnegative delays $\tau_{ij} \ge 0$ (we say, in this case, the equilibrium is *absolutely globally attractive*); (ii) this equilibrium remains to be the global attractor for all nonnegative but small delays $\tau_{ij} \ge 0$ (we say, in this case, small delay is harmless)?

As mentioned at the beginning of this section, in the application to CAM it is desirable for the network to possess multiple stable equilibria. Assuming multiple equilibria exist and some are asymptotically stable in the absence of delay, then the same problems of "absolutely global attractivity" and "small harmless delay" are important in the practical design of artificial neural networks. Again, little has been done for these problems except for the case of monotone networks where generic convergence can be granted no matter whether delay is considered or not. To be more precise, consider (5.2.1) where each signal function is bounded and $f_j'(x_j) > 0$ for all $1 \le j \le n$, $x_j \in \mathbb{R}$. Assume that a change of variables has already been performed so that $w_{ij} \ge 0$, also assume that $\tau_{ij} = \tau_j$ (delay depends on the source node only) for all $1 \le i, j \le n$. Then similarly to the discussions in Section 4.4, one can show that system (5.2.1) generates a monotone (order-preserving) semiflow in the phase space

$$C([-\tau_1, 0]; \mathbb{R}) \times \cdots \times C([-\tau_n, 0]; \mathbb{R}).$$

Furthermore, if W is irreducible then this semiflow is eventually strongly monotone (strongly order-preserving) (Smith [1987]). Consequently, the monotone dynamical systems theory mentioned in Section 4.4 can be applied to conclude, roughly speaking, that the delay does not change the stability of a given equilibrium [5] and does not alter the generic convergence of the network. See also Olien and Bélair [1997]. In particular, if (5.2.1) has only one equilibrium, then this equilibrium is always globally attractive.

---

[5] Note that the delay does not change the set of equilibria of system (5.2.1).

5.3 Delay-induced instability

Comparing with the little progress towards the global dynamics analysis, considerable developments have been made in our understanding of the impact of the signal delay on the local stability of a given equilibrium. In this section, we consider the local stability of the zero solution $x_i(t) = 0$, $1 \le i \le n$, of the following system of delay differential equations

$$\dot{x}_i(t) = -x_i(t) + \sum_{j=1}^{n} w_{ij} f[x_j(t - \tau)], \quad 1 \le i \le n, \tag{5.3.1}$$

with the signal function $f : \mathbb{R} \to \mathbb{R}$ being continuously differentiable and $f(0) = 0$. We should point out that if we are interested in the stability of a non-zero equilibrium x^*, we can make a change of variable $y = x - x^*$ and consider the stability of the zero solution $y_i(t) = 0$, $1 \le i \le n$, of the new system for y.

Note that we assumed that all signal functions and signal delays are the same for the different neurons. The delay-induced instability problem for (5.3.1) with multiple delays is very difficult and very little has been done.

Recall that the standard phase space is

$$C = C([-\tau, 0]; \mathbb{R}^n)$$

and we use $x(\varphi)$ to denote the solution of (5.3.1) subject to the initial condition $x_0 = \varphi \in C$. We say the zero solution is *stable* if for every $\varepsilon > 0$ there exists $\delta > 0$ so that $\|\varphi\| < \delta$ implies $\|x_t(\varphi)\| < \varepsilon$ for all $t \ge 0$; *asymptotically stable* if the zero solution is stable and if there exists $\delta_0 > 0$ so that $\|\varphi\| < \delta_0$ implies $x_t(\varphi) \to 0$ as $t \to \infty$. We say that the zero solution is *unstable* if it is not stable.

The above stability properties are closely related to those of the so-called *linearization* of (5.3.1) at the zero solution, which is given by the linear system of delay differential equations

$$\dot{y}_i(t) = -y_i(t) + \beta \sum_{j=1}^{n} w_{ij} y_j(t - \tau) \tag{5.3.2}$$

with

$$\beta = f'(0) \tag{5.3.3}$$

denoting the *neural gain*. Let $W = [w_{ij}]$ be the connection matrix and I the $n \times n$ identity matrix. It is known (see, for example, Hale [1977]) that the stability of the zero solution of (5.3.2) is intimately related to the distribution of the zeros (called *characteristic values*) of the following *characteristic equation*

$$\det[\lambda I + I - e^{-\lambda \tau} \beta W] = 0 \tag{5.3.4}$$

obtained by substituting $y(t) = e^{\lambda t} c$ into (5.3.2) for a nonzero vector $c \in \mathbb{R}^n$. The following result holds (see, Hale [1977]).

Lemma 5.3.1.

   (i) *The zero solution of (5.3.2) is asymptotically stable if and only if all zeros of (5.3.4) have negative real parts;*

   (ii) *If the zero solution of (5.3.2) is asymptotically stable, then so is the zero solution of (5.3.1);*

   (iii) *If there exists a zero of (5.3.4) with positive real part, then the zero solution of (5.3.2) and the zero solution of (5.3.1) are unstable.*

It is therefore crucial to describe the distribution of all characteristic values. The following result, due to Bélair [1993], shows that these values are closely associated with the eigenvalues of the matrix W.

Lemma 5.3.2. λ *is a characteristic value if and only if there exists an eigenvalue d of the matrix W so that*

$$d\beta = (1 + \lambda)e^{\lambda \tau}. \tag{5.3.5}$$

Proof. This is trivial if we rewrite (5.3.4) as

$$\det[(\lambda + 1)e^{\lambda \tau} I - \beta W] = 0. \tag{5.3.6}$$

In view of the above result, it is important to understand the zeros of the following transcendental equation

$$(z + \alpha)e^z + \gamma = 0 \tag{5.3.7}$$

for some constants α and γ. In case both α and γ are real constants, the region of stability in the plane of the parameters (α, γ) was completely described in the following result (see Hale [1977]).

Lemma 5.3.3. *All roots of equation (5.3.7) with real constants α and γ have negative real parts if and only if*

$$\begin{cases} \alpha > -1, \\ \alpha + \gamma > 0, \\ \rho \sin \rho - \alpha \cos \rho > \gamma, \end{cases} \tag{5.3.8}$$

where $\rho = \frac{\pi}{2}$ if $\alpha = 0$, or ρ is the unique root of $\rho = -\alpha \tan \rho$ in $(0, \pi)$ if $\alpha \neq 0$. At least one of the zeros of (5.3.7) has positive real part if the $>$ sign in one of the three inequalities (5.3.8) is replaced by $<$.

To relate (5.3.5) to (5.3.7), we let $z = \lambda \tau$ to get $d\beta = \left(1 + \frac{z}{\tau}\right)e^z$, or $\tau d\beta = (\tau + z)e^z$. In other words, (5.3.5) can be written as (5.3.7) with $\alpha = \tau$ and $\gamma = -\tau d\beta$ after the change of variable $z = \lambda \tau$. It is easy to verify that the set of conditions

(5.3.8) in Lemma 5.3.3 is equivalent to $d\beta < 1$ and $\frac{1}{\cos\rho} < d\beta$ with $\rho = -\tau\tan\rho$ and $\rho \in \left(\frac{\pi}{2}, \pi\right)$, which is equivalent to

$$\text{either } d\beta \in [-1, 1)$$

$$\text{or } d\beta < -1 \text{ and } \tau < \frac{\arccos\frac{1}{d\beta}}{d\beta\sqrt{1-\left(\frac{1}{d\beta}\right)^2}} = \frac{\arccos\frac{1}{d\beta}}{\sqrt{(d\beta)^2-1}}.$$

Consequently, if we let $\sigma(W)$ denote the set of all eigenvalues of W, then

Theorem 5.3.4. *Assume that $\sigma(W) \subseteq \mathbb{R}$ and $\tau \geq 0$. Then the zero solution of system (5.3.2) is asymptotically stable if and only if, for every $d \in \sigma(W)$, either*

(i) *$\beta d \in [-1, 1)$, or*

(ii) *$\beta d < -1$ and $\tau < \dfrac{\arccos\frac{1}{d\beta}}{\sqrt{(d\beta)^2-1}}$, where the inverse cosine takes the value in the interval $\left[\frac{\pi}{2}, \pi\right]$.*

The limiting cases of arbitrarily small and arbitrarily large values of the delay τ are considered in the following result.

Corollary 5.3.5. *Assume that $\sigma(W) \subseteq \mathbb{R}$. Then*

(i) *the zero solution of (5.3.2) is asymptotically stable for $\tau = 0$ if and only if $d < \frac{1}{\beta}$ for each $d \in \sigma(W)$;*

(ii) *the zero solution of (5.3.2) is asymptotically stable for all nonnegative value of the delay τ if and only if $d \in \left[-\frac{1}{\beta}, \frac{1}{\beta}\right)$ for each $d \in \sigma(W)$.*

In the applications, it is important to assess the destabilizing influences of the presence of the delays in the governing equations. We say *delay-induced instability* is possible in (5.3.2) whenever the zero solution is asymptotically stable when $\tau = 0$ and there exists a value τ for which the zero solution is not asymptotically stable. In view of Corollary 5.3.5, we then have the following

Corollary 5.3.6. *Assume that $\sigma(W) \subseteq \mathbb{R}$. Then a necessary and sufficient condition for delay-induced instability to be possible in (5.3.2) is that*

$$-\lambda_{\min} > \frac{1}{\beta} > \lambda_{\max} \tag{5.3.9}$$

where $\lambda_{\min}$ and $\lambda_{\max}$ denote, respectively, the eigenvalues of the matrix W with minimum and maximum values.

It is more problematic to perform a stability analysis for the case of a general connection matrix W with complex eigenvalues. Nevertheless, we have the following

Lemma 5.3.7. *Assume that the number b is complex in the equation*

$$\lambda = -1 + be^{-\lambda\tau}. \tag{5.3.10}$$

Then

    (i) *all zeros of (5.3.10) have negative real parts for $\tau = 0$ if and only if* Re$(b) < 1$;

    (ii) *all zeros of (5.3.10) have negative real parts for all nonnegative values of the delay τ if and only if $|b| \leq 1$ and $b \neq 1$, where $|\cdot|$ denotes the norm of a complex number.*

Proof. (i) can be easily proved. To prove (ii), write $\sigma = \lambda + 1$ in equation (5.3.10) so that

$$\sigma e^{\sigma\tau} = be^{\tau}. \tag{5.3.11}$$

If we further let $\sigma = r + is = \rho e^{i\theta}$ and $b = Re^{i\phi}$ with $0 \leq \phi, \theta < 2\pi$, then substituting in (5.3.11) and separating real and imaginary parts yields

$$R = \rho e^{(r-1)\tau}, \tag{5.3.12}$$

$$\phi = \theta + s\tau + 2\pi j, \tag{5.3.13}$$

where j is an integer chosen so that $0 \leq \phi, \theta < 2\pi$.

Assume $R = |b| < 1$. If $r \geq 1$ then by (5.3.12) we have $\rho < 1$, which is a contradiction since $\rho = \sqrt{r^2 + s^2} \geq r$. Therefore, $R = |b| < 1$ implies $r < 1$, and thus Re $\lambda = r - 1 < 0$. If $R = |b| = 1$ and $b \neq 1$, then (5.3.12) implies $\rho e^{(r-1)\tau} = 1$. This, together with $\rho \geq r$, implies either $r < 1$ or $\rho = r = 1$. In the latter case, we have $s = 0$ and hence $\theta = 0$, $\phi = 0$ and $b = Re^{i\phi} = 1$, a contradiction. This proves the sufficiency.

To show the necessity, we fix b so that $|b| = R > 1$. Choose $s = \sqrt{R^2 - 1}$, $\theta = \arctan\sqrt{R^2 - 1} \in (0, \frac{\pi}{2})$, $\tau = \frac{\phi - \theta + 2\pi}{s}$ and $\lambda = is$. Then at the above values of τ and λ, both sides of (5.3.10) are

$$i\sqrt{R^2 - 1} = -1 + Re^{i\phi}\, e^{-i(\phi - \theta + 2\pi)} = -1 + Re^{i\theta}.$$

This shows that for the given τ, (5.3.10) has a zero of zero real part. This completes the proof.

Remark 5.3.8. The work of Bélair [1993] suggested that for a fixed $\tau > 0$, all zeros of (5.3.10) have negative real parts if and only if b lies inside the domain bounded by the arcs of Archimedean spirals $R = (\phi - \frac{\pi}{2})/\tau$ and $R = (\frac{3\pi}{2} - \phi)/\tau$ for $\frac{\pi}{2} < \phi < \pi$. Such a domain, called the zone of stability, is a teardrop shaped region, and approaches the half plane Re$(b) < 1$ in the limit as $\tau \to 0^+$, and the circle of radius 1 as $\tau \to \infty$. See Figure 5.3.1.

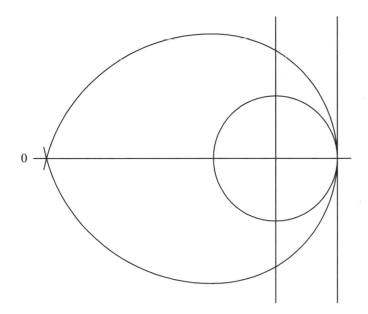

Figure 5.3.1. A teardrop region as the zone of stability of $\lambda + 1 = be^{-\lambda\tau}$ for a fixed $\tau > 0$ and its limiting regions: a half plane when $\tau \to 0$ and the unit circle when $\tau \to \infty$.

The above discussion shows that it is possible for the delay τ to destabilize an equilibrium which is asymptotically stable in the absence of time-delay, and a necessary and sufficient condition for the occurrence of delay-induced instability of the zero solution of (5.3.2) in terms of $\sigma(W)$ can be stated as

$$s(W) < \frac{1}{\beta} < \rho(W), \tag{5.3.14}$$

where

$$s(W) = \max\{\operatorname{Re}\lambda; \lambda \in \sigma(W)\},$$
$$\rho(W) = \max\{|\lambda|; \lambda \in \sigma(W)\}.$$

By contrast, if $s(W) > 1/\beta$, then the zero solution is unstable for all τ; if $1/\beta \geq \rho(W)$ and $\frac{1}{\beta} \notin \sigma(W)$ then the zero solution is asymptotically stable for all τ.

Recall that we say matrix W has the *Perron property* if $\rho(W) \in \sigma(W)$, that is, $\rho(W) = s(W)$. Therefore, equation (5.3.14) can be interpreted as (see Bélair, Campbell and van den Driessche [1996]) *Delay-induced instability for the zero solution of equation (5.3.2) is impossible for any choice of the neuron gain if and only if the connection matrix W has the Perron property.*

Recall that *directed graph* (digraph) of a real $n \times n$ matrix $W = (w_{ij})$ consists of a set of vertices $\{1, 2, \ldots, n\}$ and a set of directed edges, with edge set $\{\overrightarrow{e}_{ji}; w_{ij} \neq 0\}$.

A *cycle* of length k in this graph is a set of edges $\{\vec{e}_{i_1i_2}\}, \{\vec{e}_{i_2i_3}\}, \ldots, \{\vec{e}_{i_ki_1}\}$ with vertices $i_1, \ldots, i_k$ distinct. The *signed digraph* of W, $D(W)$, is obtained from the digraph by attaching to each edge (j, i) the sign of w_{ij}. The cycle is positive (respectively negative) according as the product $w_{i_2i_1} w_{i_3i_2} \ldots w_{i_1i_k}$ is positive (respectively negative). A network is called *frustrated* if the signed digraph of its connection matrix W has a negative cycle. In Bélair, Campbell and van den Driessche [1996], it is proved that *Delay-induced instability of the zero solution of equation* (5.3.2) *is possible only if the network is frustrated.*

This result shows that frustration is essential for delay-induced instability, confirming the numerical observation in Marcus and Westervelt [1988, 1989].

Example 5.3.9. Let

$$W = \begin{pmatrix} 0 & 1 & 0 \\ -a & 0 & 1-a \\ 0 & 1 & 0 \end{pmatrix}, \tag{5.3.15}$$

where $0 < a < 1$, be the connection matrix of a frustrated 3-neuron system. For general a, $\sigma(W) = \{0, \pm\sqrt{1-2a}\}$. When $a = 0.5$, $\sigma(W) = \{0\}$ and thus W has the Perron property for this value of a, and the zero solution of equation (5.3.2) is linearly asymptotically stable for all delays. When $a = 0.25$, $\sigma(W) = \{0, \pm\sqrt{2}/2\}$ and thus for all τ, when $\beta < \sqrt{2}$, the zero solution of equation (5.3.2) is linearly asymptotically stable, and when $\beta > \sqrt{2}$, it is unstable. By contrast, when $a = 0.75$, $\sigma(W) = \{0, \pm i\sqrt{2}/2\}$, thus W does not have the Perron property, and delay-induced instability of the zero solution of equation (5.3.2) occurs for $\beta > \sqrt{2}$ at some value of $\tau > 0$.

A network in which every cycle in the connection matrix is negative is called *fully frustrated*. The example below gives a fully frustrated network.

Example 5.3.10. Let the signed digraph of the $n \times n$ network connection matrix W consist of a single neuron (labeled 1) connected by negative 2-cycle to each of $n - 1$ other neurons (labeled 2, $\ldots$, n). The product of each 2-cycle, namely, $w_{1k}w_{k1}$, $k = 2, \ldots, n$ is negative and $\sigma(W) = \{0, \pm i \sum_{k=2}^{n} |w_{1k}w_{k1}|\}$. This is a fully frustrated network and exhibits delay-induced instability for $\beta > 1/\sum_{k=2}^{n} |w_{1k}w_{k1}|$.

When the connection matrix W is normalized as $\sum_{j=1}^{n} |w_{ij}| = 1$ for all $i = 1, \ldots, n$, Bélair, Campbell and van den Driessche [1996] obtained the following quantitative result for a network that is not frustrated:

Theorem 5.3.11. *Assume that W is normalized and is not frustrated. Then the zero solution of* (5.3.2) *is linearly asymptotically stable for all delays if $\beta < 1$, and is unstable for all delays if $\beta > 1$.*

Example 5.3.12. Theorem 5.3.11 can be applied to the all-excitatory connection matrix with $w_{ij} = 1/(n - 1)$ for $i \neq j$ and $w_{ii} = 0$. In fact, Wu [1998] proved, for the nonlinear equation (5.3.2), that the zero solution is a global attractor when $\beta < 1$, whereas two nonzero asymptotically stable equilibria appear if $\beta > 1$.

5.4 Hopf bifurcation of periodic solutions

The discussions in Section 5.3 show how delay-induced instability occurs in a network of neurons through the change of the signs of the real parts of some characteristic values of the linearization at a given equilibrium when the delay passes through a critical value. Typically, such a change of the signs of the real part of a characteristic value is related to the occurrence of a pair of purely imaginary characteristic values and, as we illustrate in this section, such a change leads to the occurrence of a branch of periodic solutions — the so-called Hopf bifurcation.

For the sake of simplicity, we will first concentrate on the simple scalar equation

$$\dot{x}(t) = -x(t) + F(x(t - \tau)) \tag{5.4.1}$$

with the time delay $\tau > 0$ as bifurcation parameter, where $F \in C^3(\mathbb{R}; \mathbb{R})$ and $F(0) = 0$. Such an equation can be regarded as the model for a single neuron with self-feedback. Alternatively, such an equation describes the synchronized activity of a group of neurons described by the additive model (5.3.1). While the signal function is usually assumed to be a smooth sigmoid which changes its concavity at zero, we emphasize that in many applications it is important not to assume that $F''(0) = 0$. This is specially useful when we consider the dynamics of the model equation

$$\dot{y}(t) = -y(t) + f(y(t - \tau)) + I$$

near an equilibrium y^* such that $y^* = f(y^*) + I$, where I is an input of constant value. A change of variable $x = y - y^*$ leads to a scalar equation of (5.4.1) with $F(\xi) = f(\xi + y^*) - y^* + I$ for $\xi \in \mathbb{R}$. It is then obvious that $F''(0) = f''(y^*)$ may be nonzero even if f is a sigmoid and even if $f''(0) = 0$.

We first rescale the time t by $t \to t/\tau$. This provides the normalization of the delay so that

$$\dot{x}(t) = -\tau x(t) + \tau F(x(t - 1)). \tag{5.4.2}$$

Let $\beta = F'(0)$. Then the linearization at the equilibrium $x = 0$ is

$$\dot{x}(t) = -\tau x(t) + \tau \beta x(t - 1) \tag{5.4.3}$$

and the associated characteristic equation is

$$0 = \Delta(z; \tau, \beta) = z + \tau - \tau \beta e^{-z}. \tag{5.4.4}$$

As illustrated in last section, the analysis of the distribution of zeros of the transcendental function in (5.4.4) yields information about the stability of the equilibrium $x = 0$.

In this section, we consider the critical values of τ such that (5.4.4) has a pair of purely imaginary zeros.

Let $z = i\omega$ with $\omega > 0$ and substitute this into (5.4.4). We have

$$i\omega + \tau - \tau\beta(\cos\omega - i\sin\omega) = 0,$$

from which it follows that

$$1 = \beta\cos\omega \qquad\qquad (5.4.5)$$

and

$$\omega = -\tau\beta\sin\omega. \qquad\qquad (5.4.6)$$

Consequently,

$$\omega = \omega_k(\beta) = \begin{cases} 2k\pi + \arccos\frac{1}{\beta} & \text{for } \beta < -1, \\ 2(k+1)\pi - \arccos\frac{1}{\beta} & \text{for } \beta > 1 \end{cases} \qquad (5.4.7)$$

and

$$\tau = \tau_k(\beta) = \frac{\omega_k(\beta)}{\sqrt{\beta^2 - 1}} \qquad \text{for } |\beta| > 1 \qquad (5.4.8)$$

for a nonnegative integer k (denoted by $k \in \mathbb{N}_0$). This proves the following

Lemma 5.4.1. *The characteristic equation* (5.4.4) *has a pair of purely imaginary conjugate complex solutions $z = \pm i\omega$ with $\omega > 0$ if and only if $|\beta| > 1$ and $\omega = \omega_k(\beta)$ and $\tau = \tau_k(\beta)$ for some $k \in \mathbb{N}_0$. The functions ω_k and τ_k are C^1-smooth and ω_k satisfies*

$$\omega_k \in \left((2k + \tfrac{3}{2})\pi, (2k+2)\pi \right) \quad \text{for } k \in \mathbb{N}_0 \text{ and } \beta > 1,$$
$$\omega_k(\beta) \in \left((2k + \tfrac{1}{2})\pi, (2k+1)\pi \right) \quad \text{for } k \in \mathbb{N}_0 \text{ and } \beta < -1.$$

τ_k is decreasing in $\beta > 1$ and increasing in $\beta < -1$ for each fixed $k \in \mathbb{N}_0$. Furthermore,

$$\tau_{k+1}(\beta) - \tau_k(\beta) = \frac{2\pi}{\sqrt{\beta^2 - 1}} \quad \text{for } \beta \in \mathbb{R} \setminus [-1, 1] \text{ and } k \in \mathbb{N}_0$$

and

$$\lim_{\beta \to \pm\infty} \tau_k(\beta) = 0, \qquad \lim_{\beta \to \pm 1} \tau_k(\beta) = \infty.$$

To apply the standard Hopf bifurcation theory, one needs to check the so-called "transversality condition" formulated in the next result.

Lemma 5.4.2. *Let $|\beta| > 1$ and fix $k \in \mathbb{N}_0$. Then there is an $\varepsilon > 0$ and a simple characteristic root $z(\tau) = a(\tau) + ib(\tau)$ of* (5.4.4) *for $|\tau - \tau_k(\beta)| < \varepsilon$ such that z is C^1 in τ, $a(\tau_k(\beta)) = 0$, $b(\tau_k(\beta)) = \omega_k(\beta)$ and $\frac{d}{d\tau}a(\tau_k(\beta)) > 0$.*

Proof. The existence of ε and $z(\tau)$ follow from the implicit function theorem since $\frac{\partial}{\partial z}\Delta(z; \tau, \beta)|_{z=i\omega_k(\beta), \tau=\tau_k(\beta)} = 1 + \tau\beta e^{-z}|_{z=i\omega_k(\beta), \tau=\tau_k(\beta)} \neq 0.$ Substituting $z(\tau)$ into (5.4.4) and differentiating with respect to τ, we get

$$z'(\tau) + 1 - \beta e^{-z(\tau)} + \tau\beta e^{-z(\tau)}z'(\tau) = 0,$$

from which and $\tau_k\beta e^{-i\omega_k} = i\omega_k + \tau_k$ (here we write $\tau_k = \tau_k(\beta)$ and $\omega_k = \omega_k(\beta)$) it follows that

$$a'(\tau_k) = \operatorname{Re}\frac{i\omega_k}{(1 + i\omega_k + \tau_k)\tau_k} = \frac{\omega_k^2}{\tau_k\left[(1 + \tau_k)^2 + \omega_k^2\right]} > 0.$$

From the results in Lemmas 5.4.1 and 5.4.2 and using some detailed analysis, from XI.2 in Diekmann, van Gils, Verduyn Lunel and Walther [1995], we can have a complete picture about the distribution of zeros of (5.4.4) shown in Figure 5.4.1. Namely, one observes that the characteristic equation (5.4.4) has

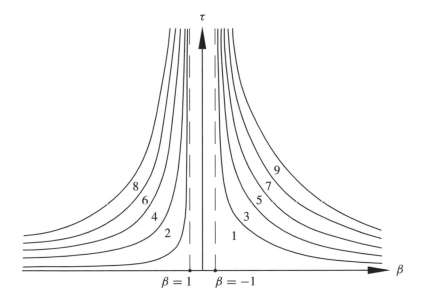

Figure 5.4.1. Schematic picture of the distribution partition of the (β, τ)-plane. Integer numbers in each region bounded by $\tau = \tau_k(\beta)$ shows the number of characteristic values with positive real parts.

(i) only solutions with negative real parts if $\tau \geq 0$ and $\beta \in [-1, 1]$ or if $\beta < -1$ and $0 \leq \tau < \tau_0(\beta)$ (so the equilibrium $x = 0$ is asymptotically stable);

(ii) exactly $2k + 1$ solutions with positive real parts if $\beta > 1$ and $\tau_{k-1}(\beta) < \tau \leq \tau_k(\beta)$ for $k \geq 1$, and exactly one solution with positive real part if $\beta > 1$ and $0 \leq \tau \leq \tau_0(\beta)$;

(iii) exactly $2k$ solutions with positive real parts provided that $\tau_{k-1}(\beta) < \tau \leq \tau_k(\beta)$
       with $\beta < -1$ and $k \in \mathbb{N}$ (the set of positive integers).

In other words, $\tau_k(\beta)$ is the sequence of critical values of the delay where the "unstable
dimension" of the equilibrium $x = 0$ increases by 2.

Lemmas 5.4.1 and 5.4.2 also provide the verification of the conditions required in
the standard Hopf bifurcation of delay differential equations (see, Hale and Verduyn
Lunel [1993], p. 332) and thus we have

Theorem 5.4.3. *Let $|\beta| > 1$. Then for each $k \in \mathbb{N}_0$, a Hopf bifurcation for equation
(5.4.1) occurs at $\tau_k(\beta)$. In particular, there exists a sequence $\{\tau^n\}_{n=1}^{\infty}$ such that
$\tau^n \to \tau_k(\beta)$ as $n \to \infty$ and that for each $\tau = \tau^n$, equation (5.4.1) has a periodic
solution x_{k,τ^n} of the period p_{k,τ^n} such that $p_{k,\tau^n} \to \frac{2\pi \tau_k(\beta)}{\omega_k(\beta)} = \frac{2\pi}{\sqrt{\beta^2-1}}$ and $x_{k,\tau^n}(t) \to$
0 uniformly for $t \to \mathbb{R}$ as $n \to \infty$.*

It is desirable to determine the direction of the above Hopf bifurcation, *i.e.*, to
answer the question whether the bifurcating branch of periodic solutions exists locally
for $\tau > \tau_k(\beta)$ (*supercritical bifurcation*) or $\tau < \tau_k(\beta)$ (*subcritical bifurcation*). Also,
it is important to describe the stability of the bifurcating periodic solutions. A general
approach to the solutions of the above questions is to compute and analyze the reduced
system on the *center manifold* associated with the conjugate pair of characteristic
values $\pm i \omega_k(\beta)$, an alternative approach is that based on the method of Lyapunov–
Schmidt developed in Stech [1985]. Two general algorithms are available in the
literature to compute the center manifold.

One is provided by Bélair and Campbell with a Maple implementation and another
is provided in Faria and Magalhães [1995] to calculate the reduced system on the center
manifold in *normal forms*. Giannakopoulos and Zapp [1999] applied the algorithm
of Faria and Maghalhães and derived the following normal form equation, in polar
coordinates (ρ, ξ), for the reduced system on the center manifold

$$\begin{cases} \dot{\rho} = (\tau - \tau_k)\frac{d}{d\tau}a(\tau_k)\rho + K\rho^2 + O((\tau - \tau_k)^2\rho + |(\rho, \tau - \tau_k)|^4), \\ \dot{\xi} = -\omega_k + O(|(\tau - \tau_k, \rho)|) \end{cases} \tag{5.4.9}$$

with

$$K = \frac{F'''(0)\tau}{2} \frac{1 + \beta^2\tau}{\beta(1 + 2\tau + \beta^2\tau^2)}$$
$$- \frac{(F''(0))^2\tau}{2} \frac{\beta^2\tau(11\beta^2 + 6\beta - 2) + (2\beta^3 + 13\beta^2 + 4\beta - 4)}{\beta^2(5\beta + 4)(\beta - 1)(1 + 2\tau + \beta^2\tau^2)} \quad \text{if } \beta < -1$$

and

$$K = \frac{F'''(0)\tau}{2} \frac{1 + \beta^2\tau}{\beta(1 + 2\tau + \beta^2\tau^2)}$$
$$- \frac{(F''(0))^2\tau}{2}$$
$$\cdot \frac{\beta^2\tau(11\beta^3 + 35\beta^2 + 24\beta - 6) + (-2\beta^4 + 11\beta^3 + 43\beta^2 + 24\beta - 12)}{\beta^2(\beta - 1)(5\beta^2 + 15\beta + 12)(1 + 2\tau + \beta^2\tau^2)}$$

if $\beta > 1$.

Since $\frac{d}{d\tau}a(\tau_k) > 0$, the sign of the factor K determines the direction and (orbitally asymptotical) stability of the bifurcating periodic solutions. Note also that

$$\pm K < 0 \quad \text{if and only if} \quad \pm F'''(0)F'(0) < \pm(F''(0))^2 C_k(\beta)$$

with

$$C_k(\beta) = \begin{cases} \frac{11\beta^2 + 6\beta - 2}{(5\beta + 4)(\beta - 1)} + 2\frac{(\beta + 1)^2}{(5\beta + 4)(1 + \beta^2\tau_k)} & \text{for } \beta < -1, \\ \frac{11\beta^3 + 35\beta^2 + 24\beta - 6}{(\beta - 1)(5\beta^2 + 15\beta + 12)} - 2\frac{(\beta + 1)(\beta^2 - 3)}{(5\beta^2 + 15\beta + 12)(1 + \beta^2\tau_k)} & \text{for } \beta > 1. \end{cases}$$

We then get

Theorem 5.4.4. *Fix $k \in \mathbb{N}$.*

(i) *If $F'''(0)F'(0) > (F''(0))^2 C_k(\beta)$, then the Hopf bifurcation is subcritical and the bifurcating periodic solutions are unstable.*

(ii) *If $F'''(0)F'(0) < (F''(0))^2 C_k(\beta)$, then the Hopf bifurcation is supercritical. The bifurcating periodic solutions are stable if $\beta < -1$ and $k = 0$, and unstable if $\beta < -1$ and $k \geq 1$ or if $\beta > 1$.*

From the definition of $C_k(\beta)$, we can easily verify that

(C1) if $\beta < -1$ or $\beta > \sqrt{3}$, then the sequence $\{C_k(\beta)\}_{k \in \mathbb{N}_0}$ is positive, bounded and strictly increasing;

(C2) if $1 < \beta < \sqrt{3}$, the sequence $\{C_k(\beta)\}_{k \in \mathbb{N}_0}$ is positive, bounded and strictly decreasing;

(C3) if $\beta = \sqrt{3}$, then the sequence $\{C_k(\beta)\}_{k \in \mathbb{N}_0}$ is a positive constant sequence.

Consequently, we have the following

Corollary 5.4.5. *Let $[F''(0)]^2 + [F'''(0)]^2 \neq 0$.*

(i) *If $F'''(0)F'(0) \leq 0$ then all Hopf bifurcations are supercritical;*

(ii) *If $F''(0) = 0$ and $F'''(0)F'(0) > 0$ then all Hopf bifurcations are subcritical;*

(iii) *If $F''(0) \neq 0$ and $F'''(0)F'(0) > 0$ then*

    (a) *for $\beta \in (-\infty, -1) \cup (\sqrt{3}, \infty)$, either all Hopf bifurcations are supercritical, or all Hopf bifurcations are subcritical, or there exists a unique $k_c \in \mathbb{N}$ such that all Hopf bifurcations occurring at $\tau = \tau_k(\beta)$ are subcritical for $k < k_c$ and supercritical for $k > k_c$;*

    (b) *for $\beta \in (1, \sqrt{3})$, either all Hopf bifurcations are supercritical, or all Hopf bifurcations are subcritical, or there exists a unique $k_c \in \mathbb{N}$ such that all Hopf bifurcations occurring at $\tau = \tau_k(\beta)$ are subcritical for $k > k_c$ and supercritical for $k < k_c$.*

Figure 5.4.2 lists all possible bifurcation scenarios.

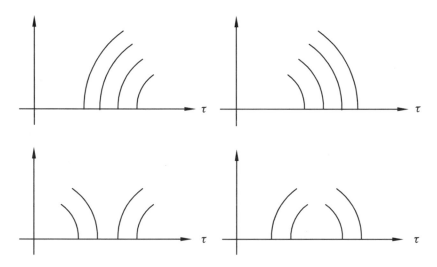

Figure 5.4.2. All possible bifurcation scenarios if $[F''(0)]^2 + [F'''(0)]^2 \neq 0$.

So far the analysis in Sections 5.3 and 5.4 is local in nature: the asymptotic stability of the equilibrium and the existence of a local Hopf bifurcation come from the study of the characteristic values of the linearization of the network at considered equilibrium and the conclusion about the behaviors holds true only for trajectories in a small neighborhood of the equilibrium. In particular, the local Hopf bifurcation theorem (Theorem 5.4.3) gives the existence of periodic solutions near $(0, \tau_k(\beta))$ in the space $C([-1, 0]; \mathbb{R}) \times \mathbb{R}$. This is a local result in the sense that (i) the amplitude of the periodic solution is small and (ii) the existence of a periodic solution is guaranteed only for τ near $\tau_k(\beta)$. The issue of global existence (for all $\tau > \tau_k(\beta)$) of large-amplitude periodic solutions is related to the study of the global continuation which we briefly describe below. Let

$$S = \{(\varphi, \tau, p) \in C([-1, 0]; \mathbb{R}) \times \mathbb{R}_+ \times \mathbb{R}_+; x \text{ is a non-constant}$$
$$p\text{-periodic solution of (5.4.2) with } x_0 = \varphi\}$$
$$\overline{S} = \text{ the closure of } S \text{ in } C([-1, 0]; \mathbb{R}) \times \mathbb{R}_+ \times \mathbb{R}_+.$$

For each $k \in \mathbb{N}_0$, let S_k denote the maximal connected component of $\overline{S}$ in $C([-1, 0]; \mathbb{R}) \times \mathbb{R}_+ \times \mathbb{R}_+$ which contains $(0, \tau_k(\beta), 2\pi/\sqrt{\beta^2 - 1})$. Assuming $F \in C^2(\mathbb{R}, \mathbb{R})$ with $|F'(0)| > 1$ and assuming that $F(\bar{x}) = \bar{x}$ for some $\bar{x} \in \mathbb{R}$ implies $F'(\bar{x}) \neq 1$, Giannakopoulos and Zapp [1998] applied the global Hopf bifurcation theorem in Erbe, Geba, Krawcewicz and Wu [1992] based on the S^1-equivariant degree theory to obtain the following:

Theorem 5.4.6. *For each $k \in \mathbb{N}_0$, the maximal connected component S_k is unbounded. Moreover, for each $(\varphi, \tau, \beta) \in S_k$ with $\varphi \neq 0$ there exists a non-constant p-periodic solution x of (5.4.1) with $x_0 = \varphi$ and $p > 2\tau$ if $\beta < -1$ and $k = 0$, $p \in \left(\frac{\tau}{k+\frac{1}{2}}, \frac{\tau}{k}\right)$ if $\beta < -1$ and $k \geq 1$ and $p \in \left(\frac{\tau}{k+1}, \frac{\tau}{k+\frac{1}{2}}\right)$ if $\beta > 1$ and $k \in \mathbb{N}$.*

To ensure the unboundedness from above of the τ-component of S_k (this is equivalent to ensuring the existence of a non-constant periodic solution for each $\tau > \tau_k(\beta)$), we need further conditions to guarantee certain *a-priori* bounds for the super-norm norm of all possible periodic solutions. This can be achieved if we assume that

there exists a compact interval $I \subseteq \mathbb{R}$ with $0 \in \text{int}(I)$ and $F(I) \subseteq I$[6]   (5.4.10)

and assume the negative feedback condition

$$x F(x) < 0 \quad \text{for } x \in I \setminus \{0\}.$$ (5.4.11)

More precisely, one has

Corollary 5.4.7. *Assume (5.4.10) holds. Then*

(i) *if $\beta < -1$, then for each $k \in \mathbb{N}$ and $\tau > \tau_k(\beta)$ there exists $\varphi \in C([-1, 0]; I)$ and $p \in \left(\frac{\tau}{k+\frac{1}{2}}, \frac{\tau}{k}\right)$ such that $(\varphi, \tau, p) \in S_k$;*

(ii) *if $\beta > 1$, then for each $k \in \mathbb{N}_0$ and $\tau > \tau_k(\beta)$ there exists $\varphi \in C([-1, 0]; I)$ and $p \in \left(\frac{\tau}{k+1}, \frac{\tau}{k+\frac{1}{2}}\right)$ such that $(\varphi, \tau, p) \in S_k$;*

(iii) *if $\beta < -1$ and, in addition, (5.4.11) holds, then for each $\tau > \tau_0(\beta)$ there exists $\varphi \in C([-1, 0]; I)$ and $p > 2\tau$ such that $(\varphi, \tau, p) \in S_k$.*

---

[6]Note that a commonly used signal function which is bounded satisfies this condition.

We now provide several examples to illustrate the above local and global results. All figures below are schematic. For the bifurcation diagrams based on a numerical continuation method for delay differential equations where the periodic solutions are approximated by Fourier polynomials, see Giannakopoulos and Zapp [1998].

Example 5.4.8. We consider the sigmoid nonlinearity

$$F(x) = F_1(x) := \frac{3}{2}\frac{1}{1 + \exp(-4x)} - \frac{3}{4}.$$

Clearly, $-\frac{3}{4} < F(x) < \frac{3}{4}$ for all $x \in \mathbb{R}$, F is strictly increasing and has exactly one turning point at 0 ($F''(0) = 0$, $F'''(0) < 0$). Moreover, $F(0) = 0$, $\beta = F'(0) = \frac{3}{2} > 1$ and F has three fixed points $\bar{x} = 0$, $\tilde{x} \approx 0.644$, $-\tilde{x}$. By Theorem 5.4.4, we conclude that a supercritical Hopf bifurcation occurs at $(0, \tau_k(\lambda))$ for each $k \in \mathbb{N}_0$. Corollary 5.4.7 shows that for each fixed $k \in \mathbb{N}_0$ and for all $\tau > \tau_k(\lambda)$ there is a periodic solution x belonging to S_k (in the sense explained in Corollary 5.4.7). Furthermore, we know that the period p of x satisfies $p \in \left(\frac{\tau}{k+1}, \frac{\tau}{k+\frac{1}{2}}\right)$.

Figures 5.4.3 and 5.4.4 show that along the branches $|x| = \sup_{t \in \mathbb{R}} |x(t)|$ is monotonically increasing in τ, but p/τ does not change much.

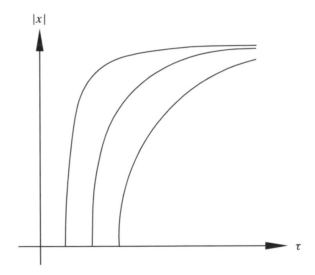

Figure 5.4.3. Schematic picture of branches of periodic solutions in the $(\tau, |x|)$-plane. All bifurcations are supercritical and the corresponding solutions are bounded by $\tilde{x}$.

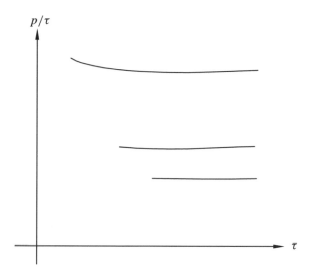

Figure 5.4.4. Schematic picture of branches of periodic solutions in the $\left(\frac{p}{\tau}, \tau\right)$-plane.

Example 5.4.9. We now use the nonlinearity $F(x) = -F_1(x)$ with F_1 as in the first example. $\bar{x} = 0$ is the only fixed point of F, and $\beta = F'(0) = -\frac{3}{2} < -1$, $F''(0) = 0$, $F'''(0) > 0$. Therefore, a supercritical Hopf bifurcation occurs at $\tau = \tau_k(\lambda)$, $k \in \mathbb{N}_0$. Since F satisfies (5.4.10) and (5.4.11) with $I = [-\tilde{x}, \tilde{x}]$, where $\tilde{x}$ is defined in Example 5.4.8, we obtain that the τ-component of S_k is unbounded from above. The period p of the solution x belonging to S_k is larger than 2τ (the corresponding solution is said to be slowly oscillating) for $k = 0$ and $p \in \left(\frac{\tau}{k+\frac{1}{2}}, \frac{\tau}{k}\right)$ for $k \geq 1$. Numerical simulations indicate that the branches are monotonically increasing in τ and for every $\tau > \tau_0(\lambda)$ there is a unique slowly oscillating periodic solution.

Example 5.4.10. We now provide an example which demonstrates the change of the direction of Hopf bifurcations. Let

$$F(x) = -575 \arctan(x + 10\sqrt{5}) + 575 \arctan(10\sqrt{5}).$$

Then $F(0) = 0$, $\beta := F'(0) = -\frac{575}{1+500} < -1$, $F''(0) = \frac{11500}{251001}\sqrt{5}$, $F'''(0) = -\frac{1723850}{125751501}$. We have $1.49716 \approx C_0(\beta) < C_1(\beta) \approx 1.49952$. We can show that a Hopf bifurcation occurs at $\tau = \tau_k(\beta)$ for each $k \in \mathbb{N}_0$. The first Hopf bifurcation at $\tau = \tau_0(\beta)$ is subcritical, all other Hopf bifurcations are supercritical (see Figure 5.4.5).

Example 5.4.11. We now consider an example where the function F does not satisfy condition (5.4.10). Consider

$$F(x) = -15 \sinh(x).$$

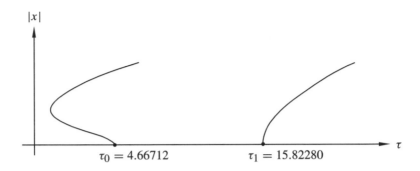

Figure 5.4.5. A subcritical Hopf bifurcation occurs at $\tau = \tau_0(\beta) = 4.66712$ and a supercritical Hopf bifurcation occurs at $\tau = \tau_1(\beta) = 15.82280$.

It holds that $F(0) = 0$, $\beta := F'(0) = -15$, $F''(0) = 0$ and $F'''(0) = -15$. F is strictly decreasing but does not satisfy condition (5.4.10). $\bar{x} = 0$ is the only fixed point of F. Using Theorem 5.4.4 we get a subcritical Hopf bifurcation at each $\tau = \tau_k(\beta)$. From Theorem 5.4.6 we know that the branches of periodic solutions S_k bifurcating from $(\bar{x} = 0, \tau = \tau_k(\beta))$ are unbounded but we do not know whether the component of S_k is unbounded. Numerically simulations show that the branches are bounded in τ. The numerically computed p-periodic solution $x_k(t; \tau)$ belonging to S_k exists only for $0 < \tau < \tau_k(\beta)$ and $|x_k| \to \infty$ as $\tau \to 0$. Furthermore, for the period p of $x_0(t; \tau)$ it was observed that $\frac{p}{\tau} \to 4$ as $\tau \to 0$.

Example 5.4.12. The last example deals with the case in which F does not satisfy the negative feedback condition (5.4.11). We consider

$$F(x) = 2\frac{x+1}{1 + (x+1)^{10}} - 2.$$

This equation was introduced by Mackey and Glass [1977] as a mathematical model for the generation of white blood cells. F is a bounded function with $F(0) = 0$, $\beta = F'(0) = -4 < -1$, $F''(0) = -5$ and $F'''(0) = 255$. Using Theorem 5.4.4 we get a supercritical Hopf bifurcation at $(\bar{x} = 0, \bar{\tau} = \tau_k(\beta))$ for every $k \in \mathbb{N}_0$. From Corollary 5.4.7 we know that for $k \geq 1$ the τ-component of S_k is unbounded from above. Since (5.4.11) is not satisfied, we can not investigate S_0 by applying Theorem 5.3.7 directly. Numerical simulations show that the branch S_0 of slowly oscillating periodic solutions undergoes a cascade of period doubling bifurcations. Moreover, numerical calculations indicate the existence of chaotic attractors for sufficiently large delay τ.

We note that the general procedure described here for the existence of a local bifurcation and its global continuation of a scalar equation can, in principle, be extended to networks of arbitrarily large number of neurons, but the practical implementation of this procedure is far from trivial and it is quite useful to use the algebraic manipulation language Maple to perform the calculations. In Bélair, Campbell and van der Driessche [1996], this was used to perform a center manifold reduction for a frustrated network of 3 neurons given by

$$\dot{u}_i(t) = -u_i(t) + \sum_{j=1}^{3} w_{ij} f(u_j(t - \tau)), \quad 1 \le i \le 3$$

with

$$W = (w_{ij}) = \begin{pmatrix} 0 & 1 & 0 \\ -1.25 & 0 & 1 \\ 1.25 & 1 & 0 \end{pmatrix}$$

and

$$f(u) = \beta \tanh(u).$$

Here, $\sigma(W) = \{1, -0.5 \pm i\}$ and delay-induced instability occurs for $1 > \beta > \frac{2}{\sqrt{5}}$, giving rise to a Hopf bifurcation. Specially, for $\beta = 0.96$, this occurs at $\tau = 4.265$ where the linearization has a pair of purely imaginary characteristic values $\pm i\omega$ with $\omega = 0.3899$. Thus at this point, the solutions to the system of 3 delay differential equations may be approximated by the flow on a two-dimensional center manifold governed by a system of two ordinary differential equations whose normal form in polar coordinates is

$$\begin{cases} \dot{\rho} = -0.541\rho^3, \\ \dot{\theta} = 0.3899. \end{cases}$$

This shows that the bifurcation is supercritical and the delay-induced stable periodic solutions occur for $\tau > 4.265$.

Finally, we mention that a general symmetric Hopf bifurcation theory was developed and applied to obtain the existence of periodic solutions with some spatial-temporal patterns for networks of neurons with delayed feedback and symmetric connection topology in Wu [1998]. See also Huang and Wu [2001] and Wu, Faria and Huang [1999] for various waves induced by signal delays.

5.5   A network of two neurons: McCulloch–Pitts nonlinearity

In the subsequent sections, we consider a system of two neurons connected to each other. Denoting the activation of the neurons by x_1 and x_2, their decay rates by γ_1 and γ_2, their connection weights by w_1 and w_2, the signal delays related to each connection by τ_1 and τ_2 and the constant inputs received by each neuron by I_1 and I_2, respectively (shown in Figure 5.5.1).

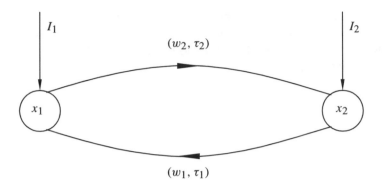

Figure 5.5.1. A network of two neurons with inputs (I_1, I_2), connection weights (w_1, w_2) and signal delays (τ_1, τ_2).

Let f denote the signal function, then we have the following system of delay differential equations

$$\begin{cases} \dot{x}_1(t) = -\gamma_1 x_1(t) + I_1 + w_1 f(x_2(t - \tau_1)), \\ \dot{x}_2(t) = -\gamma_2 x_2(t) + I_2 + w_2 f(x_1(t - \tau_2)). \end{cases}$$

Two-neuron networks have played an important role in the analysis of delay-induced changes in the dynamics of neural networks. Despite the low number of neurons in the system, in many instances, two-neuron networks with delay display the same behavior as larger networks and many techniques developed to deal with two-neuron networks can carry over to large size networks such as those used in practical applications. Therefore, a two-neuron net can be used as prototypes to improve our understanding of the influence of delay.

In this section, we introduce the work of Huang and Wu [1999b, 2000] where a two-neuron artificial network with delayed feedback from one neuron to another is considered. Assuming the signal transmission is of a digital (McCulloch–Pitts) nature, the model then describes a combination of analog and digital signal processing in the network and takes the form of a system of delay differential equations with discontinuous nonlinearity. In what follows, we present a detailed analysis of the dynamics of the inhibitory network starting from non-oscillatory states, and show that the trajectories are either synchronized and ultimately periodic or converge to a desynchronized equilibrium. We will also show that large threshold is one of the main sources for the synchronization of the network considered.

More precisely, we consider the following model

$$\begin{cases} \dot{x} = -\mu x + f(y(t - \tau_1)), \\ \dot{y} = -\mu y + f(x(t - \tau_2)), \end{cases} \tag{5.5.1}$$

with $\mu > 0$, $\tau_1 > 0$ and $\tau_2 > 0$. We point out that in the case where f is piecewise linear, *i.e.*, $f(\xi) = \frac{1}{2}(|\xi + 1| - |\xi - 1|)$ for $\xi \in \mathbb{R}$, (5.5.1) is the simplest form of the cellular neural networks introduced by Chua and Yang [1988] which have found interesting applications in, for example, image processing of moving objects and which can be implemented by a delayed cloning-template (see Roska and Chua [1992]).

We consider the case where the signal transmission is of digital nature: a neuron is either fully active or completely inactive. Thus the signal transmission is of McCulloch–Pitts type and we have

$$f(\xi) = \begin{cases} a & \text{if } \xi > \sigma, \\ b & \text{if } \xi \leq \sigma \end{cases} \tag{5.5.2}$$

for $a, b, \sigma \in \mathbb{R}$ and $a \neq b$. Therefore, the model describes a combination of analog and digital signal processing.

We only consider the case where $b > a$ (The case where $b < a$ can be considered similarly). To simplify our presentation, let us first rescale the variables by

$$t^* = \mu t, \qquad \tau = \mu \frac{\tau_1 + \tau_2}{2},$$
$$x^*(t^*) = \frac{2\mu}{b-a} x \left(t + \frac{\tau_1}{2} \right) - \frac{b+a}{b-a}, \qquad y^*(t^*) = \frac{2\mu}{b-a} y \left(t + \frac{\tau_2}{2} \right) - \frac{b+a}{b-a},$$
$$f^*(\xi) = \frac{2}{b-a} f \left(\frac{b-a}{2\mu} \xi + \frac{b+a}{2\mu} \right) - \frac{b+a}{b-a}, \qquad \sigma^* = \frac{2\mu}{b-a} \sigma - \frac{b+a}{b-a}$$

and then drop the $*$ to get

$$\begin{cases} \dot{x} = -x + f(y(t - \tau)), \\ \dot{y} = -y + f(x(t - \tau)) \end{cases} \tag{5.5.3}$$

with

$$f(\xi) = \begin{cases} -1 & \text{if } \xi > \sigma, \\ 1 & \text{if } \xi \leq \sigma. \end{cases} \tag{5.5.4}$$

The discontinuity of f makes it difficult to apply directly the theory of dynamical systems to model equation (5.5.3). But, the simple form of (5.5.3) and (5.5.4) enables us to carry out a direct elementary analysis of the dynamics of the network due to its obvious connection with the following systems of linear nonhomogeneous ordinary differential equations

$$\begin{cases} \dot{x} = -x - 1, \\ \dot{y} = -y - 1, \end{cases} \tag{5.5.5}$$

$$\begin{cases} \dot{x} = -x - 1, \\ \dot{y} = -y + 1, \end{cases} \tag{5.5.6}$$

$$\begin{cases} \dot{x} = -x + 1, \\ \dot{y} = -y + 1, \end{cases} \tag{5.5.7}$$

$$\begin{cases} \dot{x} = -x + 1, \\ \dot{y} = -y - 1. \end{cases} \tag{5.5.8}$$

As usual, we use $X = C([-\tau, 0]; \mathbb{R}^2)$ as the phase space. Note that for each given initial value $\Phi = (\varphi, \psi)^T \in X$, one can solve system (5.5.3) on intervals $[0, \tau]$,

$[\tau, 2\tau], \ldots$ successively to obtain a unique mapping $(x^\Phi, y^\Phi)^T : [-\tau, \infty) \to \mathbb{R}^2$ such that $x^\Phi |_{[-\tau,0]} = \varphi$, $y^\Phi |_{[-\tau,0]} = \psi$, $(x^\Phi, y^\Phi)^T$ is continuous for all $t \geq 0$, differentiable and satisfies (5.5.3) for almost every $t > 0$. This gives a unique solution of (5.5.3) defined for all $t \geq -\tau$. Our goal is to determine the limiting behaviors of $(x^\Phi, y^\Phi)^T$ as $t \to \infty$ for any $\Phi \in X$, but here we shall concentrate on the case where $\varphi - \sigma$ and $\psi - \sigma$ have no sign change on $[-\tau, 0]$ (initial values are either above or below the normalized threshold σ). Namely, we consider $\Phi \in X_\sigma^{+,+} \cup X_\sigma^{+,-} \cup X_\sigma^{-,+} \cup X_\sigma^{-,-} = X_\sigma$ given by

$$X_\sigma^{\pm,\pm} = \{\Phi \in X; \; \Phi = (\varphi, \psi)^T, \varphi \in C_\sigma^\pm \text{ and } \psi \in C_\sigma^\pm\}$$

with

$$C_\sigma^+ = \left\{\varphi; \quad \begin{array}{l} \varphi : [-\tau, 0] \to [\sigma, \infty) \text{ is continuous and } \varphi(t) - \sigma \\ \text{has only finitely many zeros on } [-\tau, 0] \end{array}\right\}$$

and

$$C_\sigma^- = \left\{\varphi; \quad \begin{array}{l} \varphi : [-\tau, 0] \to (-\infty, \sigma] \text{ is continuous and } \varphi(t) - \sigma \\ \text{has only finitely many zeros on } [-\tau, 0] \end{array}\right\}.$$

Clearly, all constant initial values (except σ) are contained in X_σ. Our analysis shows that once we restrict Φ to X_σ, the behaviors of $(x^\Phi(t), y^\Phi(t))^T$ as $t \to \infty$ are completely determined by the value $(\varphi(0), \psi(0))^T$ and the size of the threshold σ. More precisely, we have

Theorem 5.5.1. *Assume that $|\sigma| < 1$, $\Phi = (\varphi, \psi)^T \in X_\sigma^{+,+} \cup X_\sigma^{-,-}$ and $\varphi(0) = \psi(0)$. Then $x^\Phi(t) = y^\Phi(t)$ for all $t \geq 0$ and $x^\Phi(t) = y^\Phi(t) = q(t)$ for all $t \geq \tau + \ln[1 + \varphi(0)] - \ln(1 + \sigma)$ (respectively, for $t \geq \tau + \ln[1 - \varphi(0)] - \ln(1 - \sigma)$) if $\Phi \in X_\sigma^{+,+}$ (respectively, $\Phi \in X_\sigma^{-,-}$), where $q : \mathbb{R} \to \mathbb{R}$ is a periodic function of the minimal period $\omega = \ln\left(\frac{2}{1+\sigma}e^\tau - 1\right) + \ln\left(\frac{2}{1-\sigma}e^\tau - 1\right)$.*

Clearly, if $\Phi = (\varphi, \psi)^T \in X$ is *synchronous* (i.e., $\varphi = \psi$) then the solution $(x^\Phi, y^\Phi)^T : [-\tau, \infty) \to \mathbb{R}^2$ is *synchronized*, that is, $x^\Phi(t) = y^\Phi(t)$ for all $t \geq 0$, due to the uniqueness of the initial value problem of (5.5.3). The above result shows that the solution $(x^\Phi, y^\Phi)^T$ of (5.5.3) is synchronized even if the initial value Φ is asynchronous but $\varphi(0) = \psi(0)$ and $(\varphi, \psi)^T \in X_\sigma^{+,+} \cup X_\sigma^{-,-}$. Obviously, a synchronized solution of (5.5.3) is characterized by the scalar equation

$$\dot{z} = -z + f(z(t - \tau)) \tag{5.5.9}$$

and Theorem 5.5.1 shows that a solution of (5.5.9) with initial value in $C_\sigma^\pm$ is eventually periodic with the minimal period $\omega = \ln\left(\frac{2}{1+\sigma}e^\tau - 1\right) + \ln\left(\frac{2}{1-\sigma}e^\tau - 1\right)$ if $|\sigma| < 1$. It is interesting to note that the periodic function q and its minimal period ω are both independent of the choice of the initial value of $\Phi \in X_\sigma^{\pm,\pm}$ with $\varphi(0) = \psi(0)$. Moreover, note that $\frac{\omega}{2\tau} \to 1$ as $\tau \to \infty$.

Theorem 5.5.2. *Let $|\sigma| < 1$ and $\Phi = (\varphi, \psi)^T \in X_\sigma$. Then $\left(x^\Phi(t), y^\Phi(t)\right)^T \to$
$(-1, 1)^T$ if either $\psi(0) > \varphi(0)$ or $\varphi(0) = \psi(0) = \sigma$ and $\Phi \in X_\sigma^{-,+}$; and
$\left(x^\Phi(t), y^\Phi(t)\right)^T \to (1, -1)^T$ if either $\psi(0) < \varphi(0)$ or $\varphi(0) = \psi(0) = \sigma$ and
$\Phi \in X_\sigma^{+,-}$ as $t \to \infty$.*

As a consequence, if either $\varphi(0) \neq \psi(0)$ or $\varphi(0) = \psi(0) = \sigma$ and $(\varphi, \psi)^T \in$
$X_\sigma^{-,+} \cup X_\sigma^{+,-}$ then the solution is asynchronous and is convergent to the desynchronized equilibrium $(-1, 1)^T$ or $(1, -1)^T$. The proof, provided below, is quite
elementary and geometrically intuitive. The idea is as follows: we divide the area
above the diagonal in the $(\varphi(0), \psi(0))^T$-plane into 3-regions: region R_I consists of
$(\varphi(0), \psi(0))^T$ with $\varphi(0) \leq \sigma$ and $\psi(0) \geq \sigma$ and we show by direct calculations
that the solution $\left(x^\Phi, y^\Phi\right)^T$ is convergent to $(-1, 1)^T$ as $t \to \infty$ if $(\varphi(0), \psi(0))^T \in$
R_I and $\Phi = (\varphi, \psi)^T \in X_\sigma^{-,+}$; region R_{II} consists of $(\varphi(0), \psi(0))^T$ with either
$\varphi(0) \geq \sigma$, $\psi(0) \geq \sigma$ and $\psi(0) \geq \varphi(0)e^\tau + e^\tau - 1$ or $\varphi(0) \leq \sigma$, $\psi(0) \leq \sigma$ and
$\varphi(0) \leq \psi(0)e^\tau - e^\tau + 1$, and we show that for some $t^* > 0$, $\left(x^\Phi(t^*), y^\Phi(t^*)\right)^T \in R_I$
and $\left(x^\Phi(t^* + \cdot), y^\Phi(t^* + \cdot)\right)^T |_{[-\tau, 0]} \in X_\sigma^{-,+}$ if $(\varphi(0), \psi(0))^T \in R_{II}$ and $\Phi =$
$(\varphi, \psi)^T \in X_\sigma^{+,+} \cup X_\sigma^{-,-}$; region R_{III} consists of those above the diagonal but not in
$R_I \cup R_{II}$, and we also show that there exists $t^* > 0$ so that $\left(x^\Phi(t^*), y^\Phi(t^*)\right)^T \in R_I$
and $\left(x^\Phi(t^* + \cdot), y^\Phi(t^* + \cdot)\right)^T |_{[-\tau, 0]} \in X_\sigma^{-,+}$ if $(\varphi(0), \psi(0))^T \in R_{III}$ and $\Phi =$
$(\varphi, \psi)^T \in X_\sigma^{+,+} \cup X_\sigma^{-,-}$. In summary, the solution $\left(x^\Phi, y^\Phi\right)^T$ of (5.5.3) starting from any point above the diagonal must enter the region R_I eventually and
$\left(x^\Phi(t^* + \cdot), y^\Phi(t^* + \cdot)\right)^T |_{[-\tau, 0]} \in X_\sigma^{-,+}$ for some $t^* \geq 0$, and therefore must be convergent to $(-1, 1)^T$ as $t \to \infty$. The region below the diagonal in the $(\varphi(0), \psi(0))^T$-
plane, can be dealt with similarly.

The next result shows that the normalized threshold σ can be regarded as a bifur-
cation parameter, and the dynamics of (5.5.3) with $|\sigma| > 1$ is completely different
from that with $|\sigma| < 1$. In particular, we notice the role of synchronization of large
threshold.

Theorem 5.5.3. *Let $|\sigma| > 1$, and let $\Phi = (\varphi, \psi)^T \in X_\sigma$. Then, as $t \to \infty$,
$\left(x^\Phi(t), y^\Phi(t)\right)^T \to (1, 1)^T$ if $\sigma > 1$, and $\left(x^\Phi(t), y^\Phi(t)\right) \to (-1, -1)^T$ if $\sigma < -1$.*

In the case where $|\sigma| = 1$, we notice that the dynamics of the network undergoes
change when the ratio $\frac{\psi(0)+1}{\varphi(0)+1}$ (if $\sigma = 1$) or $\frac{\psi(0)-1}{\varphi(0)-1}$ (if $\sigma = -1$) passes certain critical
values.

Theorem 5.5.4. *Let $\sigma = 1$ and $\Phi = (\varphi, \psi)^T \in X_\sigma$. Then*

(i) $\left(x^\Phi(t), y^\Phi(t)\right)^T \to (1, 1)^T$ *if any one of the following cases occur:* (a) $\Phi \in$
$X_\sigma^{-,-}$; (b) $\Phi \in X_\sigma^{+,+}$ and $e^{-\tau} \leq \frac{\psi(0)+1}{\varphi(0)+1} \leq e^\tau$; (c) $\Phi \in X_\sigma^{-,+}$ and $\psi(0) = 1$;
(d) $\Phi \in X_\sigma^{+,-}$ and $\phi(0) = 1$;

(ii) $\left(x^{\Phi}(t), y^{\Phi}(t)\right)^{T} \rightarrow (-1, 1)^{T}$ if either $\Phi \in X_{\sigma}^{-,+}$ and $\psi(0) \neq 1$ or $\Phi \in X_{\sigma}^{+,+}$ and $\frac{\psi(0)+1}{\varphi(0)+1} > e^{\tau}$;

(iii) $\left(x^{\Phi}(t), y^{\Phi}(t)\right)^{T} \rightarrow (1, -1)^{T}$ if either $\Phi \in X_{\sigma}^{+,-}$ and $\phi(0) \neq 1$ or $\Phi \in X_{\sigma}^{+,+}$ and $\frac{\psi(0)+1}{\varphi(0)+1} < e^{-\tau}$.

Theorem 5.5.5. *Let* $\sigma = -1$ *and* $\Phi = (\varphi, \psi)^{T} \in X_{\sigma}$. *Then*

(i) $\left(x^{\Phi}(t), y^{\Phi}(t)\right)^{T} \rightarrow (-1, -1)^{T}$ *if either* $\Phi \in X_{\sigma}^{+,+}$ *and* $[\phi(0)+1][\psi(0)+1] \neq 0$ *or* $\Phi \in X_{\sigma}^{-,-}$ *and* $e^{-\tau} < \frac{\psi(0)-1}{\varphi(0)-1} < e^{\tau}$;

(ii) $\left(x^{\Phi}(t), y^{\Phi}(t)\right)^{T} \rightarrow (-1, 1)^{T}$ *if any one of the following cases occurs:*

    (a) $\Phi \in X_{\sigma}^{-,+}$;

    (b) $\Phi \in X_{\sigma}^{-,-}$ *and* $\frac{\psi(0)-1}{\varphi(0)-1} \leq e^{-\tau}$;

    (c) $\Phi \in X_{\sigma}^{+,+}$ *and* $\phi(0) = -1$;

(iii) $\left(x^{\Phi}(t), y^{\Phi}(t)\right)^{T} \rightarrow (1, -1)^{T}$ *if any one of the following occurs:*

    (a) $\Phi \in X_{\sigma}^{+,-}$;

    (b) $\Phi \in X_{\sigma}^{-,-}$ *and* $\frac{\psi(0)-1}{\varphi(0)-1} \geq e^{\tau}$;

    (c) $\Phi \in X_{\sigma}^{+,+}$ *and* $\psi(0) = 1$.

Regarding the threshold σ as a bifurcation parameter, we show in the above results that as σ moves from the interior of interval $(-1, 1)$ to its exterior, the dynamics of model equation undergoes transition from either ultimate periodicity and synchronization or convergence to asynchronous equilibria $(1, -1)^{T}$ and $(-1, 1)^{T}$ to synchronization and convergence to synchronous equilibria $(1, 1)^{T}$ and $(-1, -1)^{T}$. These results seem to indicate that large threshold is the main source for synchronization in the network discussed.

We now provide the proofs of the above results. We first note that, by solving (5.5.3) on intervals $[0, \tau]$, $[\tau, 2\tau]$, ... successively for each $\Phi = (\varphi, \psi)^{T} \in X_{\sigma}$, the unique solution $\left(x^{\Phi}, y^{\Phi}\right)^{T} : [-\tau, \infty) \rightarrow \mathbb{R}^{2}$ satisfies

(i) $\left(x^{\Phi}, y^{\Phi}\right)^{T} : [-\tau, \infty) \rightarrow \mathbb{R}^{2}$ is continuous;

(ii) $x^{\Phi}(t)$ and $y^{\Phi}(t)$ are differentiable for all $t \geq 0$ such that
$$[x(t - \tau) - \sigma][y(t - \tau) - \sigma] \neq 0;$$

(iii) $\left(x^{\Phi}(t), y^{\Phi}(t)\right)^{T}$ satisfies (5.5.3) for all $t > 0$ with
$$[x(t - \tau) - \sigma][y(t - \tau) - \sigma] \neq 0.$$

Note also that, by the variation-of-constants formula, on the interval $[t^*, t]$, the solution of the equation

$$\dot{z} = -z - 1$$

is given by

$$z(t) = e^{-(t-t^*)}[z(t^*) + 1] - 1$$

or

$$\frac{z(t) + 1}{z(t^*) + 1} = e^{-(t-t^*)},$$

and the solution of the equation

$$\dot{z} = -z + 1$$

is given by

$$z(t) = e^{-(t-t^*)}[z(t^*) - 1] + 1$$

or

$$\frac{z(t) - 1}{z(t^*) - 1} = e^{-(t-t^*)}.$$

Consequently, each $(x^\Phi, y^\Phi)^T : [0, \infty) \to \mathbb{R}^2$ moves alternately along one of the following straight lines:

$$\frac{x(t) + 1}{y(t) + 1} = \frac{x(t^*) + 1}{y(t^*) + 1},$$

$$\frac{x(t) + 1}{y(t) - 1} = \frac{x(t^*) + 1}{y(t^*) - 1},$$

$$\frac{x(t) - 1}{y(t) - 1} = \frac{x(t^*) - 1}{y(t^*) - 1},$$

$$\frac{x(t) - 1}{y(t) + 1} = \frac{x(t^*) - 1}{y(t^*) + 1}.$$

In other words, a trajectory is a broken line. As an illustrative example, we consider $\Phi = (\varphi, \psi)^T \in X_\sigma^{+,+}$ with $|\sigma| < 1$ and $\varphi(0) < \psi(0) < e^\tau \varphi(0) + e^\tau - 1$. As the proof of Theorem 5.5.2 shows, in the particular case where $\frac{1+\sigma+2e^{2\tau}}{(3+\sigma)e^\tau} \leq \frac{\psi(0)+1}{\varphi(0)+1} < e^\tau$, the solution $(x(t), y(t))^T = (x^\Phi(t), y^\Phi(t))^T$ moves along the straight line $\frac{x(t)+1}{y(t)+1} = \frac{x(0)+1}{y(0)+1} = \frac{\varphi(0)+1}{\psi(0)+1}$ until $t_1 + \tau$, where t_1 is the first zero of $(x - \sigma)(y - \sigma)$ in $[0, \infty)$, then it moves along the line $\frac{x(t)+1}{y(t)-1} = \frac{x(t_1+\tau)+1}{y(t_1+\tau)-1}$ until $t_2 + \tau$, where t_2 is the second zero of $(x - \sigma)(y - \sigma)$ in $[0, \infty)$, and then it moves along the line $\frac{x(t)-1}{y(t)-1} = \frac{x(t_2+\tau)-1}{y(t_2+\tau)-1}$ until $t_3 + \tau$, where t_3 is the third zero of $(x - \sigma)(y - \sigma)$ in $[0, \infty)$, and then it moves towards $(-1, 1)^T$ along the line $\frac{x(t)+1}{y(t)-1} = \frac{x(t_3+\tau)+1}{y(t_3+\tau)-1}$.

Proof of Theorem 5.5.1.   We only consider the case where $\Phi = (\varphi, \psi)^T \in X_\sigma^{+,+}$. The case where $\Phi = (\varphi, \psi)^T \in X_\sigma^{-,-}$ can be dealt with analogously.

Using equation (5.5.3), we can easily obtain that $x^\Phi(t) = y^\Phi(t)$ for all $t \geq 0$. Therefore, it suffices to show that the solution x of the equation

$$\dot{x} = -x + f(x(t - \tau)) \tag{5.5.10}$$

with the initial condition $x\mid_{[-\tau,0]} = \varphi \in C_\sigma^+$ is eventually periodic with the minimal period $\ln\left(\frac{2}{1+\sigma}e^\tau - 1\right) + \ln\left(\frac{2}{1-\sigma}e^\tau - 1\right)$.

Let t_1 be the first nonnegative zero of $x - \sigma$. Then for $t \in (0, t_1 + \tau)$, except at most finitely many points, we have

$$\dot{x} = -x - 1,$$

from which and the continuity of the solution it follows that

$$x(t) = e^{-t}[\varphi(0) + 1] - 1, \quad t \in [0, t_1 + \tau]$$

and, in particular,

$$x(t_1) = e^{-t_1}[\varphi(0) + 1] - 1 = \sigma.$$

This implies

$$t_1 = \ln[\varphi(0) + 1] - \ln(1 + \sigma)$$

and

$$x(t_1 + \tau) = e^{-(t_1+\tau)}(\varphi(0) + 1) - 1 = -1 + (1+\sigma)e^{-\tau} < \sigma.$$

Also

$$x_{t_1+\tau}(\theta) := x(t_1 + \tau + \theta) = e^{-(t_1+\tau+\theta)}(\varphi(0) + 1) - 1$$

$$= (1+\sigma)e^{-(\tau+\theta)} - 1 < \sigma$$

for $\theta \in (-\tau, 0]$, which means $x_{t_1+\tau} \in C_\sigma^-$.

To move to the next step where we explicitly construct a solution of (5.5.10) beyond $[0, t_1 + \tau]$, we start with a new initial value $\varphi^* = x_{t_1+\tau}$. Let t_2 be the first zero after t_1 of $x - \sigma$. Then $t_2 > t_1 + \tau$ and on $(t_1 + \tau, t_2 + \tau)$, we have

$$\dot{x} = -x + 1$$

and hence

$$x(t) = e^{-(t-t_1-\tau)}[x(t_1 + \tau) - 1] + 1$$

$$= [\varphi(0) + 1]\left(1 - \frac{2}{1+\sigma}e^\tau\right)e^{-t} + 1$$

for $t \in [t_1 + \tau, t_2 + \tau]$. In particular,

$$x(t_2) = (\varphi(0) + 1)\left(1 - \frac{2}{1+\sigma}e^\tau\right)e^{-t_2} + 1 = \sigma$$

implies

$$t_2 = \ln(\varphi(0) + 1) + \ln\left(\frac{2}{1+\sigma}e^\tau - 1\right) - \ln(1 - \sigma).$$

Also,

$$x_{t_2+\tau}(\theta) = x(t_2 + \tau + \theta) = (\varphi(0) + 1)\left(1 - \frac{2}{1+\sigma}e^\tau\right)e^{-(t_2+\tau+\theta)} + 1$$
$$= 1 - (1 - \sigma)e^{-(\tau+\theta)} > \sigma$$

for $\theta \in (-\tau, 0]$, which means $x_{t_2+\tau} \in C_\sigma^+$.

Repeating the above arguments inductively, we can show that there exists a sequence $\{t_n\}$ ($n = 1, 2, \ldots$) such that

$$t_1 = \ln(\varphi(0) + 1) - \ln(1 + \sigma),$$

$$t_2 = \ln(\varphi(0) + 1) + \ln\left(\frac{2}{1+\sigma}e^\tau - 1\right) - \ln(1 - \sigma),$$

$$t_{2n-1} = t_{2n-3} + \ln\left(\frac{2}{1+\sigma}e^\tau - 1\right) + \ln\left(\frac{2}{1-\sigma}e^\tau - 1\right), \quad n \geq 2,$$

$$t_{2n} = t_{2n-2} + \ln\left(\frac{2}{1+\sigma}e^\tau - 1\right) + \ln\left(\frac{2}{1-\sigma}e^\tau - 1\right), \quad n \geq 2,$$

$$x(t_n) = \sigma,$$

$$x(t_{2n-1} + \tau) = -1 + (1 + \sigma)e^{-\tau},$$

$$x(t_{2n} + \tau) = 1 - (1 - \sigma)e^{-\tau},$$

$$x(t) = (\varphi(0) + 1)e^{-t} - 1, \quad t \in [0, t_1 + \tau],$$

$$x(t) = e^{-(t-t_{2n-1}-\tau)}[x(t_{2n-1} + \tau) - 1] + 1$$
$$= -(\varphi(0) + 1)\left(\frac{2}{1+\sigma}e^\tau - 1\right)^n\left(\frac{2}{1-\sigma}e^\tau - 1\right)^{n-1}e^{-t} + 1,$$
$$t \in [t_{2n-1} + \tau, t_{2n} + \tau],$$

$$x(t) = e^{-(t-t_{2n}-\tau)}[x(t_{2n} + \tau) + 1] - 1$$
$$= (\varphi(0) + 1)\left(\frac{2}{1+\sigma}e^\tau - 1\right)^n\left(\frac{2}{1-\sigma}e^\tau - 1\right)^n e^{-t} - 1,$$
$$t \in [t_{2n} + \tau, t_{2n+1} + \tau].$$

See Figure 5.5.2.

Consequently, we can see that $x(t)$ is periodic for $t \geq t_1 + \tau$, and that its minimal period is

$$\omega = t_{2n+1} - t_{2n-1}$$
$$= \left[t_{2n-1} + \ln\left(\frac{2}{1+\sigma}e^\tau - 1\right) + \ln\left(\frac{2}{1-\sigma}e^\tau - 1\right)\right] - t_{2n-1}$$
$$= \ln\left(\frac{2}{1+\sigma}e^\tau - 1\right) + \ln\left(\frac{2}{1-\sigma}e^\tau - 1\right).$$

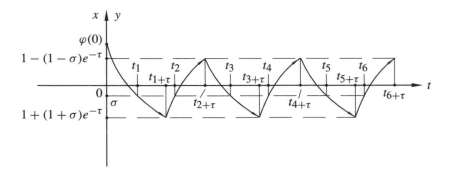

Figure 5.5.2. The solution (5.5.3) with $|\sigma| < 1$ and the initial value $(\varphi, \psi)^T \in X_\sigma^{+,+}$ and $\varphi(0) = \psi(0)$.

This completes the proof.

Proof of Theorem 5.5.2.   We only give the proof for the case of $\psi(0) > \varphi(0)$ or $\varphi(0) = \psi(0) = \sigma$ and $(\varphi, \psi)^T \in X_\sigma^{-,+}$. The conclusion for the case where $\psi(0) < \varphi(0)$ or $\varphi(0) = \psi(0) = \sigma$ and $(\varphi, \psi)^T \in X_\sigma^{+,-}$ follows immediately from the symmetry of (5.5.3) and (5.5.4).

For the sake of simplicity, we let $x(t) = x^\Phi(t)$ and $y(t) = y^\Phi(t)$ for a given $\Phi \in X_\sigma$. We distinguish several cases.

Case 1. We first consider the case where $\varphi \in C_\sigma^-$ and $\psi \in C_\sigma^+$, i.e., $\Phi = (\varphi, \psi)^T \in X_\sigma^{-,+}$. Clearly, $x(t)$ and $y(t)$ satisfy system (5.5.6) for $t \in [0, \tau]$ with $(\varphi(t - \tau) - \sigma)(\psi(t - \tau) - \sigma) \neq 0$. Therefore, for $t \in [0, \tau]$ we have

$$\begin{cases} x(t) = (\varphi(0) + 1)e^{-t} - 1, \\ y(t) = (\psi(0) - 1)e^{-t} + 1, \end{cases} \tag{5.5.11}$$

which imply that $x_\tau(\theta) = x(\tau + \theta) < \sigma$ and $y_\tau(\theta) = y(\tau + \theta) > \sigma$ for $\theta \in (-\tau, 0)$, and so $x_\tau \in C_\sigma^-$ and $y_\tau \in C_\sigma^+$. Repeating this argument on $[\tau, 2\tau]$, $[2\tau, 3\tau]$, ..., consecutively, we obtain that $x_t \in C_\sigma^-$ and $y_t \in C_\sigma^+$ for all $t \geq 0$. Therefore, $(x(t), y(t))^T$ satisfies (5.5.6) for almost all $t > 0$, from which it follows that (5.5.11) holds for all $t \in [0, \infty)$, and hence $(x(t), y(t))^T \to (-1, 1)^T$ as $t \to \infty$.

Case 2. We now consider the case where $\varphi, \psi \in C_\sigma^+$ and $\psi(0) \geq \varphi(0)e^\tau + e^\tau - 1$. The idea is simple, namely, we show that if t_1 is the first zero of $(x - \sigma)(y - \sigma)$ in $[0, \infty)$ then $x_{t_1 + \tau} \in C_\sigma^-$ and $y_{t_1 + \tau} \in C_\sigma^+$, then the conclusion follows from Case 1.

By the initial condition and (5.5.4), $(x(t), y(t))^T$ must satisfy system (5.5.5) for $t \in [0, \tau]$ except at most finitely many t. It follows that for $t \in [0, \tau]$

$$\begin{cases} x(t) = (\varphi(0) + 1)e^{-t} - 1, \\ y(t) = (\psi(0) + 1)e^{-t} - 1. \end{cases} \tag{5.5.12}$$

Let t_1 be the first zero of $(x - \sigma)(y - \sigma)$ in $[0, \infty)$. Then $(x(t), y(t))^T$ satisfies system (5.5.5) for $t \in [0, t_1 + \tau]$ except at most finitely many t, and so expression (5.5.12) holds for $t \in [0, t_1 + \tau]$.

On the other hand,

$$(x(t) - \sigma)(y(t) - \sigma) = [(\varphi(0) + 1)e^{-t} - 1 - \sigma][(\psi(0) + 1)e^{-t} - 1 - \sigma] = 0 \quad (5.5.13)$$

implies

$$t = \ln(\varphi(0) + 1) - \ln(1 + \sigma) \quad \text{or} \quad t = \ln(\psi(0) + 1) - \ln(1 + \sigma).$$

In view of $\psi(0) \geq \varphi(0)e^{\tau} + e^{\tau} - 1$, it follows that $\ln(\psi(0) + 1) \geq \ln(\varphi(0) + 1) + \tau$. Therefore,

$$t_1 = \ln(\varphi(0) + 1) - \ln(1 + \sigma) \quad \text{and} \quad t_1 + \tau \leq \ln(\psi(0) + 1) - \ln(1 + \sigma).$$

This, together with (5.5.12), implies that

$$x(t_1 + \tau) = -1 + (1 + \sigma)e^{-\tau} < \sigma,$$
$$y(t_1 + \tau) = -1 + \frac{\psi(0) + 1}{\varphi(0) + 1}(1 + \sigma)e^{-\tau} \geq \sigma,$$
$$x_{t_1+\tau}(\theta) = x(t_1 + \tau + \theta) = (\varphi(0) + 1)e^{-(t_1+\tau+\theta)} - 1$$
$$= -1 + (1 + \sigma)e^{-(\tau+\theta)} < \sigma \quad \text{for } \theta \in (-\tau, 0]$$

and

$$y_{t_1+\tau}(\theta) = y(t_1 + \tau + \theta) = (\psi(0) + 1)e^{-(t_1+\tau+\theta)} - 1$$
$$\geq -1 + (1 + \sigma)e^{-\theta} > \sigma \quad \text{for } \theta \in [-\tau, 0).$$

This shows that $x_{t_1+\tau} \in C_\sigma^-$ and $y_{t_1+\tau} \in C_\sigma^+$. So, by Case 1, we have $(x(t), y(t))^T \to (-1, 1)^T$ as $t \to \infty$.

Case 3. We next consider the case where $\varphi, \psi \in C_\sigma^-$ and $\varphi(0) \leq \psi(0)e^{\tau} - e^{\tau} + 1$. For this case, the proof is similar to that of Case 2, and thus is omitted.

Case 4. We now consider the case where $\varphi, \psi \in C_\sigma^+$ and $\varphi(0) < \psi(0) < e^{\tau}\varphi(0) + e^{\tau} - 1$. Let

$$\beta = \frac{\psi(0) + 1}{\varphi(0) + 1}.$$

Then $\varphi(0) < \psi(0) < e^{\tau}\varphi(0) + e^{\tau} - 1$ implies

$$1 < \beta < e^{\tau}.$$

Using a similar argument as that in Case 2, we can show that $(x(t), y(t))^T$ satisfies (5.5.5) for $t \in [0, t_1 + \tau]$ except at most finitely many t, where

$$t_1 = \ln[\varphi(0) + 1] - \ln(1 + \sigma),$$

and that (5.5.12) holds for all $t \in [0, t_1 + \tau]$. This, together with the fact that $1 < \beta < e^\tau$, implies that

$$x(t_1 + \tau) = -1 + (1 + \sigma)e^{-\tau} < \sigma, \quad y(t_1 + \tau) = -1 + \beta(1 + \sigma)e^{-\tau} < \sigma,$$

and

$$y(t_1) = (\psi(0) + 1)e^{-t_1} - 1 = -1 + \beta(1 + \sigma) > \sigma.$$

Therefore, there exists $t_2 \in (t_1, t_1 + \tau)$ such that $y(t_2) = \sigma$. Clearly, t_2 is the second zero of $(x - \sigma)(y - \sigma)$ in $[0, \infty)$. From (5.5.13) we can get

$$t_2 = \ln(\psi(0) + 1) - \ln(1 + \sigma) = t_1 + \ln \beta.$$

By (5.5.12), we have $x(t - \tau) < \sigma$ and $y(t - \tau) > \sigma$ for $t \in (t_1 + \tau, t_2 + \tau)$. It follows from (5.5.3) and (5.5.4) that $(x(t), y(t))^T$ satisfies system (5.5.6) for $t \in (t_1 + \tau, t_2 + \tau)$. Therefore, for $t \in [t_1 + \tau, t_2 + \tau]$, we have

$$\begin{cases} x(t) = e^{-(t - t_1 - \tau)}[x(t_1 + \tau) + 1] - 1 = (1 + \sigma)e^{-(t - t_1)} - 1, \\ y(t) = e^{-(t - t_1 - \tau)}[y(t_1 + \tau) - 1] + 1 = [(1 + \sigma)\beta - 2e^\tau]e^{-(t - t_1)} + 1. \end{cases}$$
(5.5.14)

Hence,

$$x(t_2 + \tau) = -1 + \frac{1 + \sigma}{\beta}e^{-\tau} \quad \text{and} \quad y(t_2 + \tau) = 1 - \frac{2e^\tau - (1 + \sigma)\beta}{\beta}e^{-\tau}.$$

Note that $t_2 + \tau < t_1 + 2\tau$, and that for $t \in (t_2 + \tau, t_1 + 2\tau)$ we have

$$\begin{cases} x(t - \tau) = (\varphi(0) + 1)e^{-(t - \tau)} - 1 < (\varphi(0) + 1)e^{-t_2} - 1 = -1 + \frac{1 + \sigma}{\beta} < \sigma, \\ y(t - \tau) = (\psi(0) + 1)e^{-(t - \tau)} - 1 < (\psi(0) + 1)e^{-t_2} - 1 = \sigma. \end{cases}$$

So $(x(t), y(t))^T$ must satisfy system (5.5.7) for $t \in (t_2 + \tau, t_1 + 2\tau)$. It follows that for $t \in [t_2 + \tau, t_1 + 2\tau]$ we have

$$\begin{cases} x(t) = e^{-(t - t_2 - \tau)}[x(t_2 + \tau) - 1] + 1 = [1 + \sigma - 2\beta e^\tau]e^{-(t - t_1)} + 1, \\ y(t) = e^{-(t - t_2 - \tau)}[y(t_2 + \tau) - 1] + 1 = [(1 + \sigma)\beta - 2e^\tau]e^{-(t - t_1)} + 1. \end{cases}$$
(5.5.15)

Let t_3 be the third zero of $(x - \sigma)(y - \sigma)$ in $[0, \infty)$, i.e., the first zero after t_2 of $(x - \sigma)(y - \sigma)$. Then $(x(t), y(t))^T$ satisfies system (5.5.7) for $t \in (t_2 + \tau, t_3 + \tau)$, and (5.5.15) holds for all $t \in [t_2 + \tau, t_3 + \tau]$. On the other hand, from (5.5.14), (5.5.15) and the fact $\beta > 1$, by solving the equation $(x(t) - \sigma)(y(t) - \sigma) = 0$, we can get

$$t_3 = t_1 + \ln(2e^\tau - (1 + \sigma)\beta) - \ln(1 - \sigma).$$

This, together with (5.5.15), implies that

$$x(t_3 + \tau) = 1 - \frac{2\beta e^\tau - (1 + \sigma)}{2e^\tau - (1 + \sigma)\beta}(1 - \sigma)e^{-\tau} \quad \text{and} \quad y(t_3 + \tau) = 1 - (1 - \sigma)e^{-\tau}.$$

If $\frac{1+\sigma+2e^{2\tau}}{(3+\sigma)e^{\tau}} \le \beta < e^{\tau}$, then $x(t_3+\tau) \le \sigma$, moreover, from (5.5.14) and (5.5.15), it is easy to see that $x_{t_3+\tau}(\theta) = x(t_3+\tau+\theta) < \sigma$ and $y_{t_3+\tau}(\theta) = y(t_3+\tau+\theta) > \sigma$ for $\theta \in (-\tau, 0)$. This shows that $x_{t_3+\tau} \in C_\sigma^-$ and $y_{t_3+\tau} \in C_\sigma^+$. So, by Case 1, we have $(x(t), y(t))^T \to (-1, 1)^T$ as $t \to \infty$. See Figure 5.5.3.

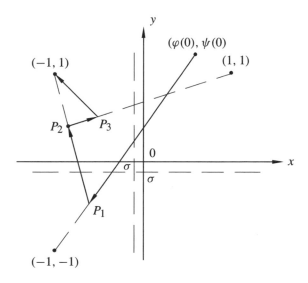

Figure 5.5.3. The trajectory of (5.5.3) with $|\sigma| < 1$ and the initial value $(\varphi, \psi)^T \in X_\sigma^{+,+}$ with $1 < \beta < e^{\tau}$ and $\frac{1+\sigma+2e^{2\tau}}{(3+\sigma)e^{\tau}} \le \beta < e^{\tau}$, where β denotes the ratio $\frac{\psi(0)+1}{\varphi(0)+1}$, P_i represents the point $(x(t_i+\tau), y(t_i+\tau))$ for $i = 1, 2, 3$.

We next consider the case where $1 < \beta < \frac{1+\sigma+2e^{2\tau}}{(3+\sigma)e^{\tau}}$. For this case, we have $x(t_3+\tau) > \sigma$. Note that $x(t_2+\tau) = -1 + \frac{1+\sigma}{\beta}e^{-\tau} < \sigma$, therefore there exists $t_4 \in (t_2+\tau, t_3+\tau)$ such that $x(t_4) = \sigma$, that is,

$$[1+\sigma - 2\beta e^{\tau}]e^{-(t_4-t_1)} + 1 = \sigma.$$

This, together with $1 < \beta < \frac{1+\sigma+2e^{2\tau}}{(3+\sigma)e^{\tau}}$, implies that

$$t_3 < t_4 = t_1 + \ln[2\beta e^{\tau} - (1+\sigma)] - \ln(1-\sigma) < t_3 + \tau.$$

Clearly, t_4 is the fourth zero of $(x-\sigma)(y-\sigma)$ in $[0, \infty)$. Again according to (5.5.14) and (5.5.15), we have $x(t-\tau) < \sigma$ and $y(t-\tau) > \sigma$ for $t \in (t_3+\tau, t_4+\tau)$. It follows from (5.5.3) and (5.5.4) that $(x(t), y(t))^T$ satisfies system (5.5.6) for $t \in (t_3+\tau, t_4+\tau)$.

Therefore, for $t \in [t_3 + \tau, t_4 + \tau]$ we have

$$
\begin{cases}
x(t) = e^{-(t-t_3-\tau)}[x(t_3+\tau)+1] - 1 \\
\qquad = \dfrac{4e^{2\tau}-4\beta e^{\tau}+1-\sigma^2}{1-\sigma}e^{-(t-t_1)} - 1, \\
y(t) = e^{-(t-t_3-\tau)}[y(t_3+\tau)-1] + 1 \\
\qquad = [(1+\sigma)\beta - 2e^{\tau}]e^{-(t-t_1)} + 1.
\end{cases}
\tag{5.5.16}
$$

In particular,

$$
x(t_4+\tau) = -1 + \frac{4e^{2\tau}-4\beta e^{\tau}+1-\sigma^2}{2\beta e^{\tau}-(1+\sigma)}e^{-\tau}
$$

and

$$
y(t_4+\tau) = 1 - \frac{2e^{\tau}-(1+\sigma)\beta}{2\beta e^{\tau}-(1+\sigma)}(1-\sigma)e^{-\tau}.
$$

Note that $t_4 + \tau < t_3 + 2\tau$, and that from (5.5.15) we have $x(t-\tau) > \sigma$ and $y(t-\tau) > \sigma$ for $t \in (t_4+\tau, t_3+2\tau)$. Consequently, $(x(t), y(t))^T$ satisfies system (5.5.5) for $t \in (t_4+\tau, t_3+2\tau)$. Therefore, for $t \in [t_4+\tau, t_3+2\tau]$ we have

$$
\begin{cases}
x(t) = e^{-(t-t_4-\tau)}[x(t_4+\tau)+1] - 1 \\
\qquad = \dfrac{4e^{2\tau}-4\beta e^{\tau}+1-\sigma^2}{1-\sigma}e^{-(t-t_1)} - 1, \\
y(t) = e^{-(t-t_4-\tau)}[y(t_4+\tau)+1] - 1 \\
\qquad = \dfrac{4\beta e^{2\tau}-4e^{\tau}+(1-\sigma^2)\beta}{1-\sigma}e^{-(t-t_1)} - 1.
\end{cases}
\tag{5.5.17}
$$

Let t_5 be the fifth zero of $(x-\sigma)(y-\sigma)$ in $[0, \infty)$, i.e., the first zero after t_4 of $(x-\sigma)(y-\sigma)$. Then $(x(t), y(t))^T$ satisfies system (5.5.5) for $t \in (t_4+\tau, t_5+\tau)$, and so (5.5.17) holds for all $t \in [t_4+\tau, t_5+\tau]$. On the other hand, from (5.5.15), (5.5.16), (5.5.17) and the fact that $\beta > 1$, by solving the equation $(x(t)-\sigma)(y(t)-\sigma) = 0$, we can obtain

$$
t_5 = t_1 + \ln[4e^{2\tau} - 4\beta e^{\tau} + 1 - \sigma^2] - \ln(1-\sigma^2).
$$

This, together with (5.5.17), implies that

$$
x(t_5+\tau) = -1 + (1+\sigma)e^{-\tau}
$$

and

$$
y(t_5+\tau) = -1 + \frac{4\beta e^{2\tau}-4e^{\tau}+(1-\sigma^2)\beta}{4e^{2\tau}-4\beta e^{\tau}+1-\sigma^2}(1+\sigma)e^{-\tau}.
$$

If $\frac{4e^{3\tau}+(5-\sigma^2)e^{\tau}}{8e^{2\tau}+1-\sigma^2} \le \beta < \frac{1+\sigma+2e^{2\tau}}{(3+\sigma)e^{\tau}}$, then $t_5 > t_3 + \tau$ and $y(t_5+\tau) \ge \sigma$. Moreover, from (5.5.16) and (5.5.17), it is easy to see that $x_{t_5+\tau}(\theta) = x(t_5 + \tau + \theta) < \sigma$ and

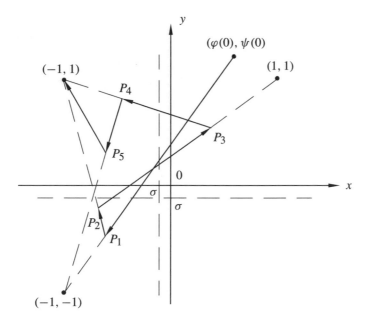

Figure 5.5.4. The trajectory of (5.5.3) with $|\sigma| < 1$ and the initial value $(\varphi, \psi)^T \in X_\sigma^{+,+}$ with $1 < \beta < e^\tau$ and $\frac{4e^{3\tau}+(5-\sigma^2)e^\tau}{8e^{2\tau}+1-\sigma^2} \le \beta < \frac{1+\sigma+2e^{2\tau}}{(3+\sigma)e^\tau}$, where β denotes the ratio $\frac{\psi(0)+1}{\varphi(0)+1}$, P_i represents the point $(x(t_i + \tau), y(t_i + \tau))$ for $i = 1, 2, 3, 4, 5$.

$y_{t_5+\tau}(\theta) = y(t_5 + \tau + \theta) > \sigma$ for $\theta \in (-\tau, 0)$. This shows that $x_{t_5+\tau} \in C_\sigma^-$ and $y_{t_5+\tau} \in C_\sigma^+$. So, by Case 1, we have $(x(t), y(t))^T \to (-1, 1)^T$ as $t \to \infty$. See Figure 5.5.4.

It now remains to consider the case where $1 < \beta < \frac{4e^{3\tau}+(5-\sigma^2)e^\tau}{8e^{2\tau}+1-\sigma^2}$. For this case we have $y(t_5 + \tau) < \sigma$. Again note that $\beta > 1$ implies

$$y(t_4 + \tau) = 1 - \frac{2e^\tau - (1+\sigma)\beta}{2\beta e^\tau - (1+\sigma)}(1-\sigma)e^{-\tau} > 1 - (1-\sigma)e^{-\tau} > \sigma.$$

Thus, there exists $t_6 \in (t_4 + \tau, t_5 + \tau)$ such that $y(t_6) = \sigma$, that is

$$y(t_6) = \frac{4\beta e^{2\tau} - 4e^\tau + \beta(1-\sigma^2)}{1-\sigma}e^{-(t_6-t_1)} - 1 = \sigma.$$

This implies that

$$t_6 = t_1 + \ln[4\beta e^{2\tau} - 4e^\tau + (1-\sigma^2)\beta] - \ln(1-\sigma^2).$$

Let

$$F(\beta) = \frac{4\beta e^{2\tau} - 4e^\tau + (1-\sigma^2)\beta}{4e^{2\tau} - 4\beta e^\tau + 1 - \sigma^2}.$$

Then

$$y(t_5 + \tau) = -1 + (1 + \sigma)F(\beta)e^{-\tau} \quad \text{and} \quad t_6 = t_5 + \ln F(\beta).$$

Note that $x(t_1 + \tau) = x(t_5 + \tau) = -1 + (1 + \sigma)e^{-\tau}$. Replacing β and t_1 by $F(\beta)$ and t_5, respectively, and applying the same procedure as above, we can obtain the following:

(I) For $t \in (t_5 + \tau, t_6 + \tau)$, $(x(t), y(t))^T$ satisfies (5.5.6). Moreover,

$$x(t) = (1 + \sigma)e^{-(t-t_5)} - 1,$$
$$y(t) = ((1 + \sigma)F(\beta) - 2e^\tau)e^{-(t-t_5)} + 1,$$
$$x(t_6 + \tau) = -1 + \frac{1+\sigma}{F(\beta)}e^{-\tau}$$

and

$$y(t_6 + \tau) = 1 - \frac{2e^\tau - (1 + \sigma)F(\beta)}{F(\beta)}e^{-\tau}.$$

(II) For $t \in (t_6 + \tau, t_7 + \tau)$, where

$$t_7 = t_5 + \ln(2e^\tau - (1 + \sigma)F(\beta)) - \ln(1 - \sigma),$$

$(x(t), y(t))^T$ satisfies (5.5.7). Moreover,

$$x(t) = (1 + \sigma - 2F(\beta)e^\tau)e^{-(t-t_5)} + 1,$$
$$y(t) = ((1 + \sigma)F(\beta) - 2e^\tau)e^{-(t-t_5)} + 1,$$
$$x(t_7 + \tau) = 1 - \frac{2F(\beta)e^\tau - (1+\sigma)}{2e^\tau - (1+\sigma)F(\beta)}(1 - \sigma)e^{-\tau}$$

and

$$y(t_7 + \tau) = 1 - (1 - \sigma)e^{-\tau}.$$

(III) If $\frac{8e^{4\tau} + (-2\sigma^2 + 8\sigma + 18)e^{2\tau} + (1+\sigma)^2(1-\sigma)}{(20+4\sigma)e^{3\tau} + (1+\sigma)(-\sigma^2 - 2\sigma + 7)e^\tau} \leq \beta < \frac{4e^{3\tau} + (5-\sigma^2)e^\tau}{8e^{2\tau} + 1 - \sigma^2}$, i.e., $\frac{1+\sigma+2e^{2\tau}}{(3+\sigma)e^\tau} \leq F(\beta) < e^\tau$, then $x(t_7 + \tau) \leq \sigma$. Moreover, $x_{t_7+\tau} \in C_\sigma^-$ and $y_{t_7+\tau} \in C_\sigma^+$. Therefore, by Case 1, we have $(x(t), y(t))^T \to (-1, 1)^T$ as $t \to \infty$. See Figure 5.5.5.

(IV) If $1 < \beta < \frac{8e^{4\tau} + (-2\sigma^2 + 8\sigma + 18)e^{2\tau} + (1+\sigma)^2(1-\sigma)}{(20+4\sigma)e^{3\tau} + (1+\sigma)(-\sigma^2 - 2\sigma + 7)e^\tau}$, i.e., $1 < F(\beta) < \frac{1+\sigma+2e^{2\tau}}{(3+\sigma)e^\tau}$, then $x(t_7 + \tau) > \sigma$ and

$$t_7 + \tau > t_8 = t_5 + \ln(2F(\beta)e^\tau - (1 + \sigma)) - \ln(1 - \sigma).$$

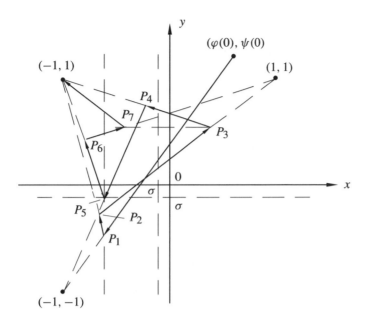

Figure 5.5.5. The trajectory of (5.5.3) with $|\sigma| < 1$ and the initial value $(\varphi, \psi)^T \in X_\sigma^{+,+}$ with $1 < \beta < e^\tau$ and $\frac{1+\sigma+2e^{2\tau}}{(3+\sigma)e^t} \le F(\beta) < e^\tau$, where β denotes the ratio $\frac{\psi(0)+1}{\varphi(0)+1}$, P_i represents the point $(x(t_i + \tau), y(t_i + \tau))$ for $i = 1, 2, \ldots, 7$.

For $t \in (t_7 + \tau, t_8 + \tau)$, $(x(t), y(t))^T$ satisfies (5.5.6). Moreover,

$$x(t) = \frac{4e^{2\tau} - 4F(\beta)e^\tau + 1 - \sigma^2}{1 - \sigma} e^{-(t-t_5)} - 1,$$

$$y(t) = [(1 + \sigma)F(\beta) - 2e^\tau]e^{-(t-t_5)} + 1,$$

$$x(t_8 + \tau) = -1 + \frac{4e^{2\tau} - 4F(\beta)e^\tau + 1 - \sigma^2}{2F(\beta)e^\tau - (1 + \sigma)} e^{-\tau},$$

and

$$y(t_8 + \tau) = 1 - \frac{2e^\tau - (1 + \sigma)F(\beta)}{2F(\beta)e^\tau - (1 + \sigma)}(1 - \sigma)e^{-\tau}.$$

(V) For $t \in (t_8 + \tau, t_9 + \tau)$, where

$$t_9 = t_5 + \ln(4e^{2\tau} - 4F(\beta)e^\tau + 1 - \sigma^2) - \ln(1 - \sigma^2),$$

$(x(t), y(t))^T$ satisfies (5.5.5). Moreover,

$$x(t) = \frac{4e^{2\tau} - 4F(\beta)e^{\tau} + 1 - \sigma^2}{1 - \sigma}e^{-(t-t_5)} - 1,$$

$$y(t) = \frac{4F(\beta)e^{2\tau} - 4e^{\tau} + (1 - \sigma^2)F(\beta)}{1 - \sigma}e^{-(t-t_5)} - 1,$$

$$x(t_9 + \tau) = -1 + (1 + \sigma)e^{-\tau},$$

and

$$y(t_9 + \tau) = -1 + (1 + \sigma)F^2(\beta)e^{-\tau},$$

where $F^2(\beta) = F(F(\beta))$.

(VI) If $\frac{16e^{5\tau}+8(7-\sigma^2)e^{3\tau}+(1-\sigma^2)(9-\sigma^2)e^{\tau}}{48e^{4\tau}+16(2-\sigma^2)e^{2\tau}+(1-\sigma^2)^2} \leq \beta < \frac{8e^{4\tau}+(-2\sigma^2+8\sigma+18)e^{2\tau}+(1+\sigma)^2(1-\sigma)}{(20+4\sigma)e^{3\tau}+(1+\sigma)(-\sigma^2-2\sigma+7)e^{\tau}}$,
i.e., $\frac{4e^{3\tau}+(5-\sigma^2)e^{\tau}}{8e^{2\tau}+1-\sigma^2} \leq F(\beta) < \frac{1+\sigma+2e^{2\tau}}{(3+\sigma)e^{\tau}}$, then $y(t_9 + \tau) \geq \sigma$, moreover, $x_{t_9+\tau} \in C_\sigma^-$
and $y_{t_9+\tau} \in C_\sigma^+$. Therefore, by Case 1, we have $(x(t), y(t))^T \to (-1, 1)^T$ as $t \to \infty$.
See Figure 5.5.6.

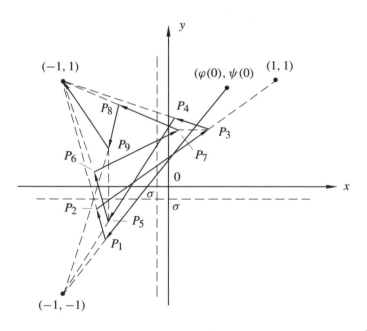

Figure 5.5.6. The trajectory of (5.5.3) with $|\sigma| < 1$ and the initial value $(\varphi, \psi)^T \in$
$X_\sigma^{+,+}$ with $1 < \beta < e^\tau$ and $\frac{4e^{3\tau}+(5-\sigma^2)e^{\tau}}{8e^{2\tau}+1-\sigma^2} \leq F(\beta) < \frac{1+\sigma+2e^{2\tau}}{(3+\sigma)e^{\tau}}$, where β
denotes the ratio $\frac{\psi(0)+1}{\varphi(0)+1}$, P_i represents the point $(x(t_i + \tau), y(t_i + \tau))$ for $i = 1,$
$2, \ldots, 9$.

(VII) If $1 < \beta < \frac{16e^{5\tau}+8(7-\sigma^2)e^{3\tau}+(1-\sigma^2)(9-\sigma^2)e^\tau}{48e^{4\tau}+16(2-\sigma^2)e^{2\tau}+(1-\sigma^2)^2}$, i.e., $1 < F(\beta) < \frac{4e^{3\tau}+(5-\sigma^2)e^\tau}{8e^{3\tau}+1-\sigma^2}$,

then

$$t_9 + \tau > t_{10} = t_9 + \ln F^2(\beta)$$

and $y(t_9 + \tau) < \sigma$. For this case, we again repeat the above procedure.

We now claim that for any $\beta \in (1, e^\tau)$, by using the above argument repeatedly, we can always find $t^* > 0$ such that

$$x_{t^*+\tau} \in C_\sigma^- \quad \text{and} \quad y_{t^*+\tau} \in C_\sigma^+. \tag{5.5.18}$$

Therefore, from Case 1, we have $(x(t), y(t))^T \to (-1, 1)^T$ as $t \to \infty$.

Suppose, by way of contradiction, that this claim is not true. Then the above procedure can be repeated inductively to generate a sequence $\{t_n\}_{n=1}^\infty$ such that t_n is the n-th zero of $(x - \sigma)(y - \sigma)$ in $[0, \infty)$. Moreover,

$$
\begin{aligned}
t_1 &= \ln(\varphi(0) + 1) - \ln(1 + \sigma), \\
t_{4n-2} &= t_{4n-3} + \ln F^{n-1}(\beta), \\
t_{4n-1} &= t_{4n-3} + \ln(2e^\tau - (1+\sigma)F^{n-1}(\beta)) - \ln(1 - \sigma), \\
t_{4n} &= t_{4n-3} + \ln(2F^{n-1}(\beta)e^\tau - (1+\sigma)) - \ln(1 - \sigma), \\
t_{4n+1} &= t_{4n-3} + \ln(4e^{2\tau} - 4F^{n-1}(\beta)e^\tau + 1 - \sigma^2) - \ln(1 - \sigma^2);
\end{aligned}
$$

$$
\begin{cases}
x(t_1 + \tau) = -1 + (1+\sigma)e^{-\tau}, \\
y(t_1 + \tau) = -1 + \beta(1+\sigma)e^{-\tau};
\end{cases}
$$

$$
\begin{cases}
x(t_{4n-2} + \tau) = -1 + \dfrac{1+\sigma}{F^{n-1}(\beta)}e^{-\tau}, \\
y(t_{4n-2} + \tau) = 1 - \dfrac{2e^\tau - (1+\sigma)F^{n-1}(\beta)}{F^{n-1}(\beta)}e^{-\tau};
\end{cases}
$$

$$
\begin{cases}
x(t_{4n-1} + \tau) = 1 - \dfrac{2F^{n-1}(\beta)e^\tau - (1+\sigma)}{2e^\tau - (1+\sigma)F^{n-1}(\beta)}(1-\sigma)e^{-\tau}, \\
y(t_{4n-1} + \tau) = 1 - (1-\sigma)e^{-\tau};
\end{cases}
$$

$$
\begin{cases}
x(t_{4n} + \tau) = -1 + \dfrac{4e^{2\tau} - 4F^{n-1}(\beta)e^\tau + 1 - \sigma^2}{2F^{n-1}(\beta)e^\tau - (1+\sigma)}e^{-\tau}, \\
y(t_{4n} + \tau) = 1 - \dfrac{2e^\tau - (1+\sigma)F^{n-1}(\beta)}{2F^{n-1}(\beta)e^\tau - (1+\sigma)}(1-\sigma)e^{-\tau};
\end{cases}
$$

$$
\begin{cases}
x(t_{4n+1} + \tau) = -1 + (1+\sigma)e^{-\tau}, \\
y(t_{4n+1} + \tau) = -1 + (1+\sigma)F^n(\beta)e^{-\tau};
\end{cases}
$$

$$
\begin{cases}
x(t) = \varphi(t), \\
y(t) = \psi(t),
\end{cases} \quad t \in [-\tau, 0];
$$

$$
\begin{cases}
x(t) = (1+\sigma)e^{-(t-t_1)} - 1, \\
y(t) = (1+\sigma)\beta e^{-(t-t_1)} - 1,
\end{cases} \quad t \in [0, t_1 + \tau];
$$

$$\begin{cases} x(t) = (1+\sigma)e^{-(t-t_1)} - 1, \\ y(t) = [(1+\sigma)\beta - 2e^\tau]e^{-(t-t_1)} + 1, \end{cases} \quad t \in [t_1 + \tau, t_2 + \tau];$$

$$\begin{cases} x(t) = [1+\sigma - 2F^{n-1}(\beta)e^\tau]e^{-(t-t_{4n-3})} + 1, \\ y(t) = [(1+\sigma)F^{n-1}(\beta) - 2e^\tau]e^{-(t-t_{4n-3})} + 1, \end{cases} \quad t \in [t_{4n-2} + \tau, t_{4n-1} + \tau];$$

$$\begin{cases} x(t) = \dfrac{4e^{2\tau} - 4F^{n-1}(\beta)e^\tau + 1 - \sigma^2}{1-\sigma}e^{-(t-t_{4n-3})} - 1, \\ y(t) = [(1+\sigma)F^{n-1}(\beta) - 2e^\tau]e^{-(t-t_{4n-3})} + 1, \end{cases} \quad t \in [t_{4n-1} + \tau, t_{4n} + \tau];$$

$$\begin{cases} x(t) = \dfrac{4e^{2\tau} - 4F^{n-1}(\beta)e^\tau + 1 - \sigma^2}{1-\sigma}e^{-(t-t_{4n-3})} - 1, \\ y(t) = \dfrac{4F^{n-1}(\beta)e^{2\tau} - 4e^\tau + (1-\sigma^2)F^{n-1}(\beta)}{1-\sigma}e^{-(t-t_{4n-3})} - 1, \end{cases} \quad t \in [t_{4n} + \tau, t_{4n+1} + \tau];$$

$$\begin{cases} x(t) = (1+\sigma)e^{-(t-t_{4(n+1)-3})} - 1, \\ y(t) = [(1+\sigma)F^n(\beta) - 2e^\tau]e^{-(t-t_{4(n+1)-3})} + 1, \end{cases} \quad t \in [t_{4n+1} + \tau, t_{4(n+1)-2} + \tau];$$

where $n = 1, 2, \ldots$, $F^0(\beta) = \beta$, $F^1(\beta) = F(\beta)$ and $F^n(\beta) = F(F^{n-1}(\beta))$ for $n \geq 2$. Furthermore, it is necessary that, for all $n = 1, 2, \ldots$,

$$x(t_{4n-1} + \tau) > \sigma, \quad i.e., \quad t_{4n-1} + \tau > t_{4n} \tag{5.5.19}$$

and

$$y(t_{4n+1} + \tau) < \sigma, \quad i.e., \quad t_{4n+1} + \tau > t_{4(n+1)-2}. \tag{5.5.20}$$

Otherwise, if n_0 is the first positive integer such that (5.5.19) or (5.5.20) does not hold, then a similar argument as above shows that (5.5.18) holds for $t^* = t_{4n_0-1}$ or $t^* = t_{4n_0+1}$, respectively, which means the above claim is true.

From the above expressions of t_n and $x(t_n + \tau)$, it is clear that (5.5.19) is equivalent to

$$F^{n-1}(\beta) < \frac{1 + \sigma + 2e^{2\tau}}{(3+\sigma)e^\tau}, \quad n = 1, 2, \ldots \tag{5.5.21}$$

and that (5.5.20) is equivalent to

$$F^n(\beta) < e^\tau, \quad n = 1, 2, \ldots. \tag{5.5.22}$$

On the other hand, after some simple calculations, we have

$$\frac{d}{d\beta}F(\beta) = \frac{[4e^{2\tau} - (1+\sigma)^2][4e^{2\tau} - (1-\sigma)^2]}{(4e^{2\tau} - 4\beta e^\tau + 1 - \sigma^2)^2} > 0 \tag{5.5.23}$$

and

$$\frac{d}{d\beta}(F(\beta) - \beta) = \frac{32\beta e^{2\tau}(e^\tau - \beta) + 16e^{2\tau}(\beta^2 - 1) + 8(1 - \sigma^2)\beta e^\tau}{(4e^{2\tau} - 4\beta e^\tau + 1 - \sigma^2)^2} > 0 \tag{5.5.24}$$

for all $1 < \beta < e^\tau$. (5.5.23) and (5.5.24) imply that $F(\beta)$ and $F(\beta) - \beta$ are strictly increasing on $(1, e^\tau)$. Since

$$\lim_{\beta \to 1+} (F(\beta) - \beta) = 0,$$

we have $F(\beta) - \beta > 0$ for all $\beta \in (1, e^\tau)$ which, together with the monotonicity of $F(\beta)$ and $F(\beta) - \beta$, implies that

$$F^n(\beta) > F^{n-1}(\beta) \quad \text{and } F^{n+1}(\beta) - F^n(\beta) > F^n(\beta) - F^{n-1}(\beta)$$

for all $\beta \in (1, e^\tau)$ and $n = 1, 2, \ldots$. Therefore, we have

$$\begin{aligned}
F^n(\beta) &= \beta + \sum_{i=1}^{n} (F^i(\beta) - F^{i-1}(\beta)) \\
&> \beta + n(F(\beta) - \beta) \\
&= \beta + n \frac{4(\beta^2 - 1)e^\tau}{4e^\tau(e^\tau - \beta) + 1 - \sigma^2}
\end{aligned}$$

for all $\beta \in (1, e^\tau)$ and $n = 1, 2, \ldots$. Note that $\frac{4(\beta^2-1)e^\tau}{4e^\tau(e^\tau-\beta)+1-\sigma^2} > 0$ for $\beta \in (1, e^\tau)$. It follows that

$$\lim_{n \to \infty} F^n(\beta) = \infty$$

for all $\beta \in (1, e^\tau)$, which contradicts (5.5.21) and (5.5.22), and hence the above claim is true.

Case 5. We finally consider the case where $\varphi, \psi \in C_\sigma^-$ and $\psi(0) > \varphi(0) > e^\tau \psi(0) - e^\tau + 1$. For this case, the proof is similar to that of Case 4, and thus is omitted. This completes the proof of Theorem 5.5.2.

The essential part in the above proof is the connection between (5.5.3) and a one-dimensional map. This will be used later in our discussions on transient behaviors.

Similar arguments are also applied to show that delay-induced oscillation can be prevented by increasing the threshold or by decreasing the synaptic connection in the following frustrated network

$$\begin{cases} \dot{x} = -x + f(y(t - \tau)), \\ \dot{y} = -y - f(x(t - \tau)), \end{cases} \tag{5.5.25}$$

with f given in (5.5.4). In particular, we establish the existence of stable limit cycle in (5.5.25) if $|\sigma| < 1$: the existence of such a cycle is independent of the size of τ though the period and the amplitude of such a cycle depend on τ. We also show that the important parameter for nonlinear oscillations is not τ but σ, and the critical values of σ are ± 1 in the sense that if $|\sigma| \geq 1$ then each trajectory of (5.5.25) converges to a single equilibrium. More precisely, we obtain

Theorem 5.5.6. *Let $|\sigma| < 1$. Then there exist $\Phi_0 = (\varphi_0, \psi_0)^T \in X_\sigma$ and $T_0 \geq 0$ such that the solution $(x^{\Phi_0}(t), y^{\Phi_0}(t))^T$ of (5.5.25) with initial value Φ_0 is periodic for $t \geq T_0$. Moreover, $\lim_{t \to \infty}[x^\Phi(t) - x^{\Phi_0}(t)] = 0$ and $\lim_{t \to \infty}[y^\Phi(t) - x^{\Phi_0}(t)] = 0$ for every solution $(x^\Phi(t), y^\Phi(t))^T$ of (5.5.25) with $\Phi = (\varphi, \psi)^T \in X_\sigma$.*

The proof is very similar to that given above for (5.5.3). The basic idea is to show that a typical trajectory of (5.5.25) is spiraling and rotates around the point (σ, σ). Figure 5.5.7 shows the periodic orbit in case $|\sigma| < 1$, this orbit can be obtained as follows:

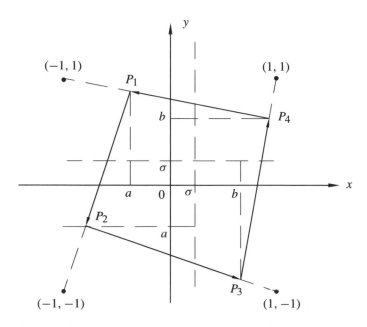

Figure 5.5.7. The periodic trajectory of (5.5.25) with $|\sigma| < 1$, where $(\varphi, \psi)^T \in X_\sigma^{+,+}$ satisfies $\frac{\psi(0)-1}{\varphi(0)+1} = \lambda_0$ for a critical value λ_0. The trajectory starts at $P_0 = (\varphi(0), \psi(0))$ and moves along four broken lines until P_4. The value $(\varphi, \psi)^T$ is chosen so that P_4 coincides with P_0 to obtain a periodic orbit. Note $a = -1 + (1+\sigma)e^{-\tau}, b = 1 - (1-\sigma)e^{-\tau}$.

Choose a point $P_0 = (\varphi(0), \psi(0))$ in the 2-dimensional Euclidean space (not the phase space) such that $(\varphi(0) - \sigma, \psi(0) - \sigma)$ is in the first quadrant and $(\varphi, \psi) \in X_\sigma^{+,+}$. The trajectory moves first along the line connecting P_0 to $(-1, 1)$ until $P_1 = (x(t_1 + \tau), y(t_1 + \tau))$, where t_1 is the first time when the trajectory meets the vertical line $x = \sigma$. Then the trajectory changes its direction and moves along the line connecting P_1 to $(-1, -1)$ until $P_2 = (x(t_2 + \tau), y(t_2 + \tau))$, where t_2 is the first time when the trajectory meets the horizontal line $y = \sigma$. In the same way, the trajectory moves to $P_3 = (x(t_3 + \tau), y(t_3 + \tau))$ and then to $P_4 = (x(t_4 + \tau), y(t_4 + \tau))$ along the line connecting P_2 to $(1, -1)$ and then the line connecting P_3 to $(1, 1)$, with t_3 and t_4 being the next zeros of $x(t) - \sigma$ and $y(t) - \sigma$, respectively. The periodic orbit is obtained by choosing $(\varphi(0), \psi(0))$ so that P_0 coincides with P_4. The existence of such $(\varphi(0), \psi(0))$ is guaranteed by the existence of a fixed point to a 1-dimensional increasing map. Figures 5.5.8 and 5.5.9 show that a typical trajectory spirals towards the above limit cycle from outside or from inside.

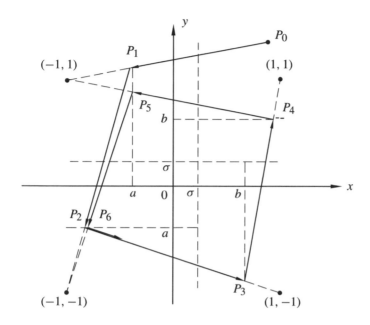

Figure 5.5.8. A typical trajectory of (5.5.25) with $|\sigma| < 1$, where the initial value $(\varphi, \psi)^T \in X_\sigma^{+,+}$ is chosen so that $\frac{\psi(0)-1}{\varphi(0)+1} > \lambda_0$. The trajectory spirals towards the periodic orbit along broken lines: from $P_0 = (\varphi(0), \psi(0))$ to P_1 along the line connecting P_0 to $(-1, 1)$, then from P_1 to P_2 along the line connecting P_1 to $(-1, -1)$ and so on so forth, here $P_i = (x(t_i + \tau), y(t_i + \tau))$ for $i = 1, 2, \ldots$ and $t_1 < t_2 < \cdots$ is the consecutive zeros of $[x(t) - \sigma][y(t) - \sigma]$.

Note that the existence of a limit cycle is independent of the size of τ provided $|\sigma| < 1$. The next result shows that increasing $|\sigma|$ is the key to prevent the occurrence of a limit cycle.

Theorem 5.5.7.   (i) *Let $|\sigma| > 1$ and $\Phi = (\varphi, \psi)^T \in X_\sigma$. Then, as $t \to \infty$, $(x^\Phi(t), y^\Phi(t))^T \to (1, -1)^T$ if $\sigma > 1$, and $(x^\Phi(t), y^\Phi(t))^T \to (-1, 1)^T$ if $\sigma < -1$;*

(ii) *Let $\sigma = 1$. Then, as $t \to \infty$, $(x^\Phi(t), y^\Phi(t))^T \to (1, -1)^T$ if either $\Phi \in X_\sigma^{+,+} \cup X_\sigma^{-,+} \cup X_\sigma^{-,-}$ or $\Phi \in X_\sigma^{+,-}$ and $\varphi(0) = 1$, and $(x^\Phi(t), y^\Phi(t))^T \to (1, 1)^T$ if $\Phi \in X_\sigma^{+,-}$ and $\varphi(0) \neq 1$;*

(iii) *Let $\sigma = -1$. Then, as $t \to \infty$, $(x^\Phi(t), y^\Phi(t))^T \to (-1, 1)^T$ if either $\Phi \in X_\sigma^{+,+} \cup X_\sigma^{+,-} \cup X_\sigma^{-,-}$ or $\Phi \in X_\sigma^{-,+}$ and $\psi(0) = -1$, and $(x^\Phi(t), y^\Phi(t))^T \to (-1, -1)^T$ if $\Phi \in X_\sigma^{-,+}$ and $\psi(0) \neq -1$.*

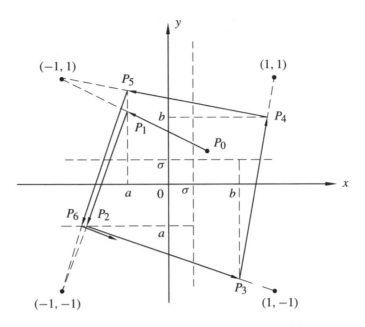

Figure 5.5.9. A typical trajectory of (5.5.25) with $|\sigma| < 1$, where the initial value $(\varphi, \psi))^T \in X_\sigma^{+,+}$ satisfies $\frac{\psi(0)-1}{\varphi(0)+1} < \lambda_0$. The trajectory spirals towards the periodic orbit along broken lines in the same way as in Figure 5.5.8, but from inside the periodic orbit.

5.6　Delay-induced transient oscillations

Numerical investigations indicate that even when precautions are taken to avoid delay-induced instabilities in the Hopfield's network of neurons, oscillations may arise in the presence of delays. These are delay-induced transient oscillations (DITOs), and cannot be predicted by the analysis of the asymptotic dynamics of neural networks. Examples of DITOs were first reported in two mutually exciting neurons and in rings by Babcock and Westervelt [1987] and Baldi and Attyia [1994]. In Pakdaman, Grotta-Ragazzo, Malta and Vibert [1997a] and Pakdaman, Grotta-Ragazzo and Malta [1998a], it was observed that when DITOs appear, the duration of the transients, that is, the time required for the system to stabilize at an equilibrium, increases as the characteristic charge-discharge time of the neurons tends to zero, or, equivalently, as the delay is increased to infinity. In fact, DITOs can last so long that for practical purposes they are indistinguishable from sustained oscillations. Therefore, understanding the origin of DITOs in neural networks constitutes an important complement to the analysis of delay induced instabilities in order to avoid long lasting oscillations that deteriorate network performance.

It seems that one can attribute the presence of DITOs to the fact that solutions of
(5.2.1) transiently follow those of the difference system defined by

$$x_i(t) = \frac{1}{\mu_i}\left[I_i + \sum_{j=1}^{n} w_{ij} f_j(x_j(t - \tau_{ij}))\right], \quad 1 \le i \le n. \tag{5.6.1}$$

System (5.6.1) can display attracting oscillations even when almost all solutions of
the corresponding delay equation (5.2.1) are convergent. The fact that some solutions
display transient oscillations before stabilizing at equilibrium is the result of the difference between the asymptotic dynamics of the continuous dynamical system (5.4.1)
and the discrete network (5.6.1).

The main purpose of this section is to confirm the above observation through the
analytical calculation of the transient regime duration (TRD) for the following two-
neuron network

$$\begin{cases} \dot{x}(t) = -x(t) + wf(y(t - \tau)), \\ \dot{y}(t) = -y(t) + wf(x(t - \tau)), \end{cases} \tag{5.6.2}$$

where f is the step function, i.e., $f(x) = 1$ if $x > 0$ and -1 otherwise.

Following Pakdaman, Grotta-Ragazzo and Malta [1998a], we show that the dynamics of system (5.6.2) can be understood in terms of the iterations of a one-dimensional
map.

We first provide a geometrical description of the construction of this map, and then
state the general results.

Assuming that the delay τ is set to zero, system (5.6.2) defines a two-dimensional
ordinary differential equation, the trajectories of which in the (x, y) plane can be easily
constructed. We denote by $r_1 = -(w, w)$, $r_2 = (0, 0)$, $r_3 = (w, w)$, $r_1' = (-w, w)$,
and $r_3' = (w, -w)$. r_1 and r_3 are the stable equilibria of the system to which most
trajectories converge. Trajectories with initial conditions (u, v) such that $u > 0$
and $v > 0$ are the segments of the straight lines connecting the initial point to r_3.
An example of such a trajectory in the two-dimensional phase space is shown in
Figure 5.6.1.

For an initial condition (u, v) with $0 \le -u < v$, the trajectory coincides with
the straight line connecting the initial point to r_3', until reaching the y-axis, that is, as
long as $x(t) < 0$. From this point on, the trajectory is the segment connecting the
intersection point with the y-axis to the equilibrium point r_3. Thus, such trajectories
are broken lines. One example is also shown in Figure 5.6.1. The trajectories of other
initial conditions can be constructed in the same way.

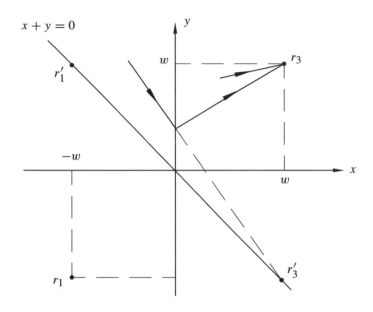

Figure 5.6.1. Two typical trajectories of the network (5.6.2) with delay $\tau = 0$.

In the following, we will mainly concentrate on the influence of delays on the trajectories of initial conditions which are constant mappings with constant values (u, v) satisfying either $u > 0$ and $v > 0$ or $0 \leq -u < v$. These two regions of initial conditions are referred to as regions I and II, respectively. Trajectories of other constant initial conditions can be derived from these through symmetry considerations.

Since the input-output function of the neurons is a step function, delays do not alter the trajectories with initial conditions in I. These are the same as for the system in the absence of delay, that is, they are segments connecting the initial condition to r_3, as exemplified in Figure 5.6.2.

The situation is different for initial conditions in II. In the case without delay, the effect of the sign change in x appeared instantaneously, as the trajectory switched to the straight line connecting r_3 right upon crossing the y-axis. When interunit signal transmission is not instantaneous, this effect is delayed, that is, the trajectory continues along the same straight line towards the point r'_3, during a time interval equal to the delay τ, before switching to the segment that takes it towards the point r_3. For all trajectories starting in region II, the break points have exactly the same abscissae. Thus, there is a critical value of v, denoted by v_1, such that the break point is exactly on the x-axis. It is easy to verify that $v_1 = w(e^\tau - 1)$. This point plays an important role, as all initial conditions in II whose trajectories cross the y-axis above v_1 reach the stable point r_3 after exactly *one* break point, such as the one shown in Figure 5.6.2.

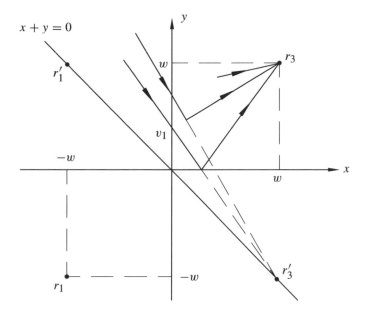

Figure 5.6.2. Some typical trajectories of the network (5.6.2) with $\tau > 0$: a trajectory starting in region I is the segment connecting the initial condition to r_3; a trajectory starting from region II moves towards r_3' first. If the intersection of the trajectory with the y-axis is above v_1 then the trajectory reaches r_3 after one break point.

Trajectories of other initial conditions will display two or more break points before they enter the region where $y < 0$, that is, at some point, the second variable changes sign. This sign change will affect the trajectory after a time delay, and provoke a second break point, where the trajectory switches to the straight line leading towards r_1'. Other break points follow this, so that eventually the broken line representing the trajectory will end up at the equilibrium point r_3. An example of a trajectory with three break points is shown in Figure 5.6.3.

The analysis hinges upon the construction of a map that describes the trajectories after the second break point. Let us first note that the break point and the trajectory with an initial condition in II are entirely determined by the ordinate v at which the trajectory first intersects the y-axis.

So in the following we only take this value into consideration. Then, for a given value v the construction of the map relies on the observation that we can find a point on the x-axis, denoted by $g(v)$ in Figure 5.6.3, whose trajectory coincides exactly with that of the original initial condition *after* the second break point. Geometrically, this point is easily obtained as the intersection point between the x-axis and the straight line connecting r_1' to the second break point of the trajectory. Thanks to this simple

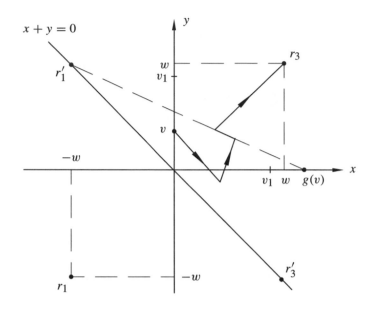

Figure 5.6.3. A trajectory with 3 break points.

geometrical construction and after elementary calculations about the values of $x(\tau+t^*)$ and $y(\tau + t^*)$ where t^* is the first time where the trajectory intersects the y-axis, we can obtain $g(v)$ analytically as

$$g(v) = \frac{w(2 + e^{-\tau})v}{2w - (v + w)e^{-\tau}}. \tag{5.6.3}$$

Once the point $(g(v), 0)$ on the x-axis is determined, we can reiterate the same analysis and construction assuming that the initial condition is in fact this new point. Thus, we can easily see that if $g(v) > v_1$, the trajectory will have only one more break point before tending to the equilibrium r_3 along a straight line. This situation is represented in Figure 5.6.3.

Conversely, when $g(v) < v_1$, the trajectory will have at least two more breaking points, since it will cross the y-axis again. For such points, we can construct a point $(0, s)$ on the y-axis such that the trajectory after the second break point, $i.e.$, the fourth break point along the trajectory starting from the initial condition, coincides with that of $(0, s)$. Thanks to the symmetries of the system, we have $s = g(g(v)) = g^2(v)$. If $g^2(v) > v_1$, we stop the process because after the fifth break point, the trajectory will converge to the stable equilibrium point along a straight line, otherwise, we continue to iterate g until there is n such that $g^n(v) > v_1$. See Figure 5.6.4. Such a value n exists because $g(v) > v$ for all $v > 0$. Figure 5.6.5 shows an example of the function g and the cobweb construction of the sequence of points $g^n(v)$, until it becomes larger than v_1.

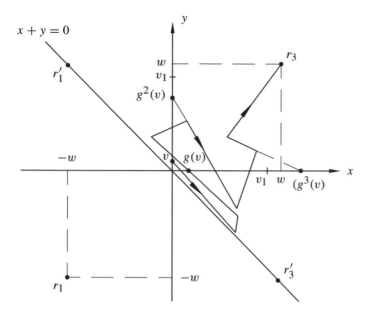

Figure 5.6.4. The sign change of y for a trajectory starting in region II provokes a second break point where the trajectory switches to the straight line leading to r'_1.

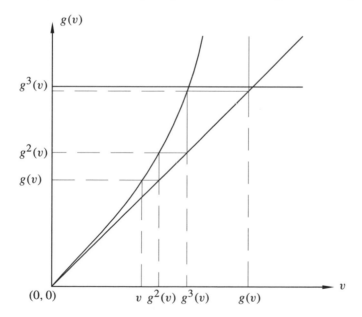

Figure 5.6.5. An example of the iteration of the map g. The curve is the graph of the function g, the horizontal line is at v_1 and the straight line is the diagonal. The cobweb construction shows the successive iterates $g(v)$ until they become larger than v_1.

We can formalize the above construction as follows:

Theorem 5.6.1. *For $r = (u, v) \in \mathbb{R}^2$ such that $v > -u \geq 0$, let $V(r) = w(v + u)/(w - u)$ and n be the integer such that $g^{n-1}(V(r)) < v_1 \leq g^n(V(r))$, where $v_1 = w(e^\tau - 1)$. Then, there exist $T \geq n\tau$ and $\theta > 0$ such that for $t \geq T$, $z(t, r) = z(t - \theta, r_n)$, where $z(t, \xi)$ is the solution of (5.6.2) through the initial condition ξ ($\xi = r$ or r_n), and r_n represents the point $(0, g^n(V(r)))$ on the y-axis for n even, and the point $(g^n(V(r)), 0)$ on the x-axis for n odd.*

Our interest is in the transient oscillations of the solutions, and the incidence of these on the TRD, that is, the time it takes the trajectory to reach a small neighborhood of the equilibrium point. Using g and its iterates we can characterize both aspects of the dynamics of the system.

Oscillations for a system like (5.6.2) can be defined as points where either one of the variables x or y take the value zero, in other words, a solution oscillates as long as there are times t such that $x(t)y(t) = 0$. We refer to such points as the zeros of the solution, and we denote by $N(r, \tau)$ the number of zeros of the solution of (5.6.2) with delay τ, going through the initial condition r. The larger the value of N, the more the solution oscillates. The following result shows that solutions of initial conditions $r = (u, v)$ in region II close to the straight line $u + v = 0$ can display arbitrarily large numbers of zeros:

Theorem 5.6.2. *Let $v_n = g^{-n}(v_1)$, then*

$$v_n = \frac{2w(2 - e^{-\tau})^n(e^\tau - 1)}{2(2 + e^{-\tau})^n + [(2 + e^{-\tau})^n - (2 - e^{-\tau})^n](e^\tau - 1)}.$$

We have $v_1(\tau) > v_2(\tau) > \cdots > v_k(\tau) > \cdots > 0$, and v_n tends to zero as $n \to \infty$. For $r = (u, v) \in \mathbb{R}^2$, with $0 \leq -u < v$, we have $N(r, \tau) = 1, 2p$ and $2p + 1$ $(p \geq 1)$ for $v > v_1 - (1 + v_1/w)u$, $v = v_p - (1 + v_p/w)u$ and $v_{p+1} - [1 + (v_{p+1})/w]u < v < v_p - (1 + v_p/w)u$, respectively.

Recall that the trajectory through $(0, v)$ is a broken line, and it converges to a stable equilibrium along a straight line after several break points. We call the time required for the trajectory to reach the starting point of the final straight line along which the trajectory converges to an equilibrium the *transient region duration*(TRD). The number of zeros of a solution informs us about the TRD, because successive zeros cannot be arbitrarily close to one another, and as long as the solution is oscillating, it is obviously not stabilizing at the equilibrium point. Thus, any measure of the length of transient oscillations provides a lower bound for the TRD. We have the following result:

Theorem 5.6.3. *We denote by $T(r, \tau)$ the transient regime duration of a solution of (5.6.2) with delay τ going through the initial condition $r = (u, v) \in \mathbb{R}^2$ with $0 \leq -u < v$. Then, $T(r, \tau) \geq p\tau$ for $v_{p+1} - [1 + (v_{p+1})/w]u < v \leq v_p - (1 + v_p/w)u$.*

This result indicates that for fixed delay τ, the TRD grows indefinitely as initial conditions get closer to the line $u + v = 0$. In Pakdaman, Grotta-Ragazzo and Vibert [1998a], a numerical calculation is carried out that shows the TRD, as a function of the ordinate v for initial conditions $(-10^{-3}, v)$, increases abruptly as v approaches 10^{-3}.

All the previous analyses were carried out for a system with fixed delay values. Our main concern is to see how the results are modified when the delay is increased. This is done in the following theorem which stems from the fact that the values $v_n(\tau)$ of the bounds delimiting the regions with a given number of zeros, and hence with a given range of TRD, increase exponentially with the delay.

Theorem 5.6.4. *For a fixed initial condition $r = (u, v)$ with $0 \leq -u < v$, there is a strictly increasing sequence $0 < \tau_1 < \tau_2 < \cdots < \tau_k < \ldots$, such that $z(t, r)$, the solution of (5.6.2) with delay τ going through r, displays exactly 1, $2p$, or $2p + 1$ zero(s) for $\tau < \tau_1$, $\tau = \tau_p$, or $\tau_p < \tau < \tau_{p+1}$, respectively. In other words, $N(r, \tau) = 1$, $2p$, and $2p + 1$ for $\tau < \tau_1$, $\tau = \tau_p$, and $\tau_p < \tau < \tau_{p+1}$, respectively.*

Thus, for fixed r, $N(r, \tau)$ and consequently $T(r, \tau)$ are increasing functions of τ, when τ is large enough. The speed with which these quantities increase can then be determined by remarking that $v_n(\tau) \approx 2we^\tau/(n + 2)$ as $\tau \to \infty$, and by using the lower bound on the TRD provided above. We have the following result on exponential increase of transient regime duration.

Theorem 5.6.5. *Let $r = (u, v) \in \mathbb{R}^2$ with $0 \leq -u < v$, and let $p = [2we^\tau/(w + V)] - 1$ where the brackets indicate the integer part, and $V = w(u + v)/(w - u)$. Then, we have $T(r, \tau) \geq p\tau$ or $T(r, \tau) \geq 2w\tau e^\tau/(w + V)$ as $\tau \to \infty$.*

Numerical simulations carried out in Pakdaman, Grotta-Ragazzo and Malta [1998a] confirmed that the above lower bound for TRD is in good agreement with the true value of TRD.

All the analysis and results presented until this point were carried out for a steplike input-output function g. This assumption allowed us to carry out a detailed analytical description of the system dynamics. Numerical investigations show that similar results hold for the system

$$\begin{cases} \dot{x}(t) = -x(t) + wg(y(t - \tau)), \\ \dot{y}(t) = -y(t) + wg(x(t - \tau)), \end{cases} \tag{5.6.4}$$

where the signal functions are $g(x) = \tanh(\alpha x)$ with $\alpha w > 1$ (Pakdaman, Grotta-Ragazzo and Malta [1998a]) and for an excitatory ring (Pakdaman, Malta, Grotta-Ragazzo, Arino and Vibert [1997c]).

5.7   Effect of delay on the basin of attraction

We now consider the following system of delay differential equations

$$
\begin{cases}
\dot{x}_1(t) = -\gamma_1 x_1(t) + I_1 + w_1 f(x_2(t - \tau_1)), \\
\dot{x}_2(t) = -\gamma_2 x_2(t) + I_2 + w_2 f(x_1(t - \tau_2))
\end{cases}
\tag{5.7.1}
$$

with

$$
f(x) = \frac{1}{1 + e^{-x}}, \qquad x \in \mathbb{R}.
\tag{5.7.2}
$$

We note that all results presented here can be extended to general smooth signal functions, but to describe some numerical simulations it is more convenient to specify the signal function involved. As illustrated before, many networks are designed so that stable equilibrium points represent stored information and transmission delay may render some equilibria unstable, thus deteriorating information retrieval. Much progress has been made to derive sufficient conditions on network parameters such as connection weights, network connection topology and signal delays that ensure that the behaviors of networks with or without delay are similar. In particular, in Sections 5.3 and 5.2, we provide two types of conditions:

(i) Constraints that ensure that the delay does not induce the loss of information stored in the stable equilibrium points. This is achieved when the system with delay has exactly the same stable equilibria as the one without delay, that is, the delay does not alter the local stability of the stable equilibrium points.

(ii) Constraints that ensure that the delay does not induce spurious stable undamped oscillations. This is achieved when both the system without delay and the one implementing delay are (almost) convergent, that is, the delay does not alter the global stability of the system.

Constraints (i) and (ii) avoid delay induced changes in the asymptotic behaviors of the system. However, even under such constraints, the delay may influence important features in the system's dynamics. For instance, changing the delay can alter the boundary of the basins of attraction of the stable equilibrium points. This can be of prime importance in associative memory networks in which the position of the basin boundaries determines which information is retrieved for a given initial condition. In this section, we are going to introduce the work of Pakdaman, Grotta-Ragazzo, Malta, Arino and Vibert [1998b] for the network (5.7.1) to show that constraints (i) and (ii) may not be sufficient to preserve the shape of the basin boundaries.

Namely, we consider the dynamics of a network of two neurons connected through delayed excitatory connections. This system has been chosen because it is simple enough to allow thorough theoretical and numerical analyses, while it retains essential features of large networks. Pakdaman *et al.* showed that this two neuron network satisfies both of the conditions stated above, that is, the delay does not affect the local stability of its stable equilibria, and, no matter what value the delays take, the two-neuron network remains almost convergent. Yet, they showed that the boundary separating the basins of attraction of two stable equilibria depends on the delays.

Clearly, we can choose the phase space $X = C([-\tau_2, 0]; \mathbb{R}) \times C([-\tau_1, 0]; \mathbb{R})$ equipped with the supremum norm. For an initial condition $\phi = (\phi_1, \phi_2) \in X$, there exists a unique solution of (5.7.1), denoted by $z(t, \phi) = (x_1(t, \phi), x_2(t, \phi))$, such that $x_1(\theta, \phi) = \phi_1(\theta)$ for $-\tau_2 \leq \theta \leq 0$, $x_2(\theta, \phi) = \phi_2(\theta)$ for $-\tau_1 \leq \theta \leq 0$ and $z(t, \phi)$ satisfies (5.7.1) for $t \geq 0$. Also for such a solution of (5.7.1), we denote by $z_t(\phi) = ((x_1)_t(\phi), (x_2)_t(\phi))$ the element of X defined by $(x_1)_t(\theta) = x_1(t + \theta)$ for $\theta \in [-\tau_2, 0]$ and $(x_2)_t(\theta) = x_2(t + \theta)$ for $\theta \in [-\tau_1, 0]$. We then obtain a semiflow $\{z_t(\varphi)\}_{t \geq 0}$ on X.

In spite of the differences between (5.7.1) and their related ODEs mentioned above, there are cases in which their dynamics are, in some sense, very similar. For (5.7.1), this happens when $w_1 w_2 > 0$ (positive feedback condition). In the rest of this section, we will assume that both connection weights are strictly positive, that is, $w_1 > 0$ and $w_2 > 0$.[7]

For $\phi = (\phi_1, \phi_2)$ and $\psi = (\psi_1, \psi_2)$ in X, we say that $\phi \geq \psi$ (resp. $\phi > \psi$) if $\phi_1(\theta') \geq \psi_1(\theta')$ (resp. $\phi_1(\theta') > \psi_1(\theta')$) for all $\theta' \in [-\tau_2, 0]$ and $\phi_2(\theta) \geq \psi_2(\theta)$ (resp. $\phi_2(\theta) > \psi_2(\theta)$) for all $\theta \in [-\tau_1, 0]$. Note that if $\phi > \psi$, the set defined by $\{\chi \in X; \phi > \chi > \psi\}$ is an open subset of X.

The positive feedback condition and Theorem 2.5 in Smith [1987] (see also discussions in Section 5.2) imply that (5.7.1) generates an eventually strongly monotone semiflow, that is,

$$\text{if } \phi \geq \phi' \text{ and } \phi \neq \phi' \text{ then } z_t(\phi) > z_t(\phi') \text{ for } t > 2\max\{\tau_1, \tau_2\}.$$

Throughout this section, constant functions in X are identified with the value they take in $\mathbb{R}^2$. Note that $z(t) = (x_1(t), x_2(t))$ taking the value (a, b), that is, $x_1(t') = a$ for $t' \geq -\tau_2$ and $x_2(t) = b$ for $t \geq -\tau_1$, is a solution of (5.7.1) if and only if (a, b) satisfies the system:

$$\begin{cases} -\gamma_1 a + I_1 + w_1 f(b) = 0, \\ -\gamma_2 b + I_2 + w_2 f(a) = 0. \end{cases} \tag{5.7.3}$$

The number and the value of the solutions of system (5.7.3) depend on the values of the parameters $(\gamma_1, \gamma_2, w_1, w_2, I_1, I_2)$. Generally, the parameter set can be separated into two regions, one in which the system has a unique solution and another in which it has three solutions. These solutions do not depend on the delays τ_1 and τ_2. Geometrically, they are the intersection points between the curves $-\gamma_1 a + I_1 + w_1 f(b) = 0$ and $-\gamma_2 b + I_2 + w_2 f(a) = 0$ in the (a, b)-plane. An example of three equilibria is shown in Figure 5.7.1, and another example of one equilibrium is shown in Figure 5.7.2.

---

[7]We point out that if $w_1 < 0$ and $w_2 < 0$ then a simple change of variables $y_1 = x_1$ and $y_2 = -x_2$ leads to a new network equation for (y_1, y_2) for which the connection weights are strictly positive.

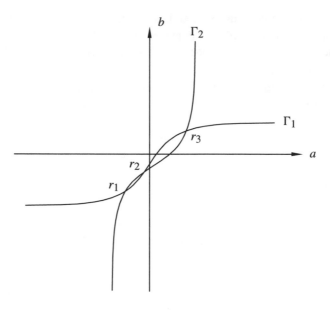

Figure 5.7.1. The case of three equilibrium points of the two-neuron system. The equilibrium points of (5.7.1) are the intersection points between the curves $\Gamma_1 : -\gamma_1 a + I_1 + w_1 f(b) = 0$ and $\Gamma_2 : -\gamma_2 b + I_2 + f(a) = 0$.

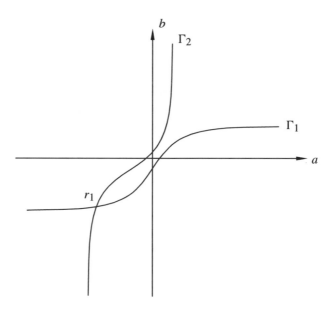

Figure 5.7.2. The case of one equilibrium point of the two-neuron system. The equilibrium point of (5.7.1) is the intersection point between the curves $\Gamma_1 : -\gamma_1 a + I_1 + w_1 f(b) = 0$ and $\Gamma_2 : -\gamma_2 b + I_2 + f(a) = 0$.

In the rest of this section we will restrict our attention to the case where system (5.7.3) has three distinct solutions. The constant functions associated with these solutions constitute the equilibria of (5.7.1) and will be denoted by r_1, r_2 and r_3, so that, when considered as constant functions in X, they are ordered as $r_1 < r_2 < r_3$.

The eventually strong monotonicity property and the results in Smith [1987] imply that the asymptotic dynamics of (5.7.1) and its related ODE (obtained by setting $\tau_1 = \tau_2 = 0$) is essentially the same in the following sense:

(P1) The equilibrium r_k, $k = 1, 2, 3$, of (5.7.1) is locally asymptotically stable if and only if the same is true of the related ODE;

(P2) The union of the basins of attraction of the stable equilibria of (5.7.1) is an open and dense set in the phase space X, the same being true for the corresponding ODE.

Recall that the basin of attraction of a stable equilibrium is the set of initial conditions in X whose trajectories converge to the equilibrium point.

From the above results it can be deduced that r_1 and r_3 are locally asymptotically stable equilibrium points, r_2 is unstable, and the union of the basins of attraction of r_1 and r_3 is an open and dense set in X. So that neither (5.7.1) nor its related ODE has stable undamped oscillatory solutions.

Our focus in this section is the effect of delay on the boundary B separating the basin of attraction of r_1 from that of r_3. Any neighborhood of a point in B intersects the basins of attraction of both r_1 and r_3. The following description of the boundary separating the two basins of attraction provides the basis for a method of numerical approximations.

Theorem 5.7.1. *Let u be a given element in X such that $u > 0$. There exists a continuous, strictly decreasing map*[8] $b_u : X \to \mathbb{R}$ *such that*

(i) *For all $\phi \in X$, $\phi + b_u(\phi)u$ is the unique intersection between the line going through ϕ and directed by u (i.e., the set $\{\phi + \lambda u; \lambda \in \mathbb{R}\}$) with the boundary separating the two basins of attraction.*

(ii) *The set $\{\phi \in X; b_u(\phi) > 0\}$ is exactly the basin of attraction of r_1.*

(iii) *The set $\{\phi \in X; b_u(\phi) < 0\}$ is exactly the basin of attraction of r_3.*

(iv) *The set $\{\phi \in X; b_u(\phi) = 0\}$ is exactly the boundary separating the two basins of attraction.*

Proof. We first prove the following claim:

The solution going through an initial condition smaller (resp. larger) than r_2 and different from r_2 converges to r_1 (resp. r_3), that is, for $\phi \in X$, if $\phi \leq r_2$ (resp. $\phi \geq r_2$) and $\phi \neq r_2$ then $z_t(\phi)$ converges to r_1 (resp. r_3) as $t \to \infty$ (see Figure 5.7.3).

---

[8]We say that $b_u : X \to \mathbb{R}$ is strictly decreasing if $\phi \geq \phi'$ and $\phi \neq \phi'$ in X implies $b_u(\phi) < b_u(\phi')$

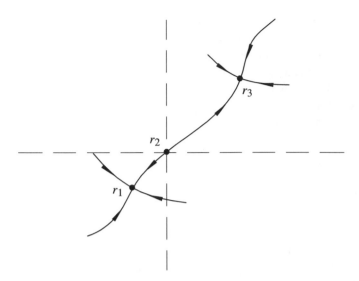

Figure 5.7.3. The convergence of a typical trajectory starting from a point below r_2 or above r_2 in the order relation of X.

To prove the claim, let $\phi \in X$ be such that $\phi \leq r_2$ and $\phi \neq r_2$. Then, from the monotonicity of the semiflow, we deduce that $z_T(\phi) < r_2$ for T sufficiently large. The set $Q = \{\chi \in X; z_T(\phi) < \chi < r_2\}$ is an open set. From property (P2) we deduce that Q intersects the union of the basins of attraction of the two stable equilibrium points: there is $\psi \in Q$ such that $z_t(\psi)$ converges to either r_1 or r_3. However, since $\psi \in Q$ we must have $z_t(\psi) < r_2$ for $t \geq 0$. Therefore, we must have $z_t(\psi) \to r_1$ as $t \to \infty$. In the same way, it can be shown that there is ψ' smaller than ϕ such that $z_t(\psi')$ tends to r_1. Therefore, for t sufficiently large, $z_t(\phi)$ is bounded by two solutions converging to r_1, so that $z_t(\phi)$ tends to r_1. In a similar way, it can be shown that for $\phi \geq r_2$ with $\phi \neq r_2$, $z_t(\phi)$ converges to r_3. This proves the claim.

Let $u \in X$ be such that $u > 0$. For a given $\phi \in X$ and a real number c, we define the translated function ϕ_c given by $\phi_c = \phi + cu$. Given $\phi \in X$, there exist real number c' and c'' such that $\phi_c < r_2$ if $c < c'$ and $\phi_c > r_2$ if $c > c''$. From the claim we deduce that, as $t \to \infty$, $z_t(\phi_c) \to r_1$ if $c < c'$ and $z_t(\phi_c) \to r_3$ if $c > c''$. Now, let us denote by $b'(\phi)$ the supremum of the set of values of c such that $z_t(\phi_c) \to r_1$ as $t \to \infty$ and by $b''(\phi)$ the infimum of the set of values of c such that $z_t(\phi_c) \to r_3$ as $t \to \infty$. Clearly, we have $b'(\phi) \leq b''(\phi)$. We claim that $b'(\phi) = b''(\phi)$. Indeed, suppose that this is false. Then, there is an open set Q of initial conditions $w \in X$ such that $\phi + b'(\phi)u < w < \phi + b''(\phi)u$. According to the monotonicity property, solutions going through initial conditions in Q converges neither to r_1 nor r_3. This implies that the open set Q does not intersect the union of the basins of attraction of

r_1 and r_3, which contradicts property (P2). So, $b'(\phi) = b''(\phi)$ and we denote this value by $b_u(\phi)$. As the basins of attraction of the two stable equilibria r_1 and r_3 are open sets, $\phi + b_u(\phi)u$ does not belong to either of the basins and is necessarily in the boundary. The characterization of the basins and the boundary described in the theorem are derived from the construction of b_u. The fact that the map b_u from X to $\mathbb{R}$ is continuous and strictly decreasing stems from the continuous dependence of the solutions of (5.7.1) on initial conditions, and from the monotonicity of the semiflow. This completes the proof.

Corollary 5.7.2. *Let ϕ belong to the boundary B, then $z_t(\psi)$ converges to r_1 (resp. r_3) as $t \to \infty$ for all $\psi \leq \phi$ (resp. $\psi \geq \phi$) such that $\psi \neq \phi$.*

Proof. Let ϕ be in the boundary. For $\psi \in X$ such that $\psi \leq \phi$ and $\psi \neq \phi$, we have $z_T(\psi) < z_T(\phi)$ for T sufficiently large. Let u be defined by $u = z_T(\phi) - z_T(\psi) > 0$. Then we have $z_T(\phi) = z_T(\psi) + u$. As $z_T(\phi)$ belongs to the boundary we have $b_u(z_T(\psi)) = 1 > 0$, so $z_T(\psi)$, and therefore, ψ belongs to the basin of attraction of r_1. In a similar way, it can be shown that $z_t(\psi)$ converges to r_3 for $\psi \geq \phi$ and $\psi \neq \phi$. This completes the proof.

From Theorem 5.7.1, we can deduce that the boundary has a regular structure. Indeed, let $u > 0$ be in X, and H be a hyperplane supplementary to u. Therefore, for all $\phi \in X$, we can write in a unique way $\phi = h + \lambda u$, where $h \in H$ and $\lambda \in \mathbb{R}$. λu is the projection of ϕ onto the line $\mathbb{R}u$ along the direction H. We denote by $p_u(\phi) = -\lambda$, the unique real number such that $\phi + p_u(\phi)u$ belongs to H.

Let ϕ belong to the boundary B, we can write $\phi = h - p_u(\phi)u$. From the definition of b_u, we know that $b_u(h)$ is the unique real number such that $h + b_u(h)u$ belongs to B, so that we have necessarily $p_u(\phi) = -b_u(h)$. From this we can deduce that the boundary is homeomorphic to the linear hyperplane H.

Corollary 5.7.3. *The map $\phi \to \phi + p_u(\phi)u$ from B to H is a homeomorphism with inverse: $h \to h + b_u(h)u$ from H to B.*

Corollary 5.7.4. *The boundary B is the graph of a map from the hyperplane H to the line $\mathbb{R}u$.*

Proof. As the map $(h, \lambda u) \to h + \lambda u$ from $H \times \mathbb{R}u$ to X is an isomorphism, we identify these two spaces. From Corollary 5.7.3, we obtain that the boundary B is the subset of $H \times \mathbb{R}u$ defined by $\{(h, b_u(h)u); h \in H\}$. In other words, B is the graph of the map L from H to $\mathbb{R}u$ defined by $L(h) = b_u(h)u$.

The basin boundary can also be characterized in terms of the dynamics of the solutions.

Theorem 5.7.5. *The boundary B between the basins of attraction of the two locally stable equilibria r_1 and r_3 of (5.7.1) is constituted by the initial value ϕ such that the solution $z_t(\phi)$ either converges to r_2 or is asymptotically periodic.*

A sketch of the above theorem is presented in Pakdaman, Grotta-Ragazzo, Malta, Arino and Vibert [1998b], see also the discussions in the next section.

The above results show that the space X is partitioned into three disjoint subsets that are positively invariant by the semiflow generated by (5.7.1): the two basins of attraction of the stable equilibrium points and the boundary separating them. In terms of the temporal evolution of solutions of (5.7.1), this means that there are only three types of asymptotical behaviors: solutions tend to r_1 or r_3 or oscillate around r_2. Solutions oscillating indefinitely around r_2 are not likely to be observed as they are unstable. Nevertheless, the presence of these oscillating solutions, as will be shown in later sections, is important for describing the global attractor and for describing transient oscillations which can be easily obtained in numerical investigations.

Theorem 5.7.1 provides a method to locate numerically some points on the boundary for a given initial condition. A given initial condition can be shifted "up" or "down", along a "positive" direction, until it crosses the boundary (Figure 5.7.4). However,

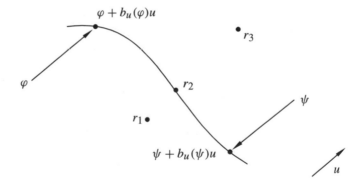

Figure 5.7.4. Schematic representation of the map b_u in a plane, where r_1, r_2 and r_3 are the equilibria and u is a strictly positive element in X. The decreasing curve is the basin boundary. All points below the boundary are in the basin of attraction of r_1, whereas all points above the boundary are in the basin of attraction of r_3. A given initial point has to be translated in the same direction as u to reach the boundary if it is in the basin of attraction of r_1, and in the opposite direction if it is in the basin of attraction of r_3. The amount by which the initial point has to be translated to reach the boundary gives the value of b_u.

carrying such a task requires extensive computations because solutions close to the boundary tend to have oscillatory transients (Babcock and Westervelt [1987]) whose duration increases exponentially with the delay (Pakdaman, Grotta-Ragazzo, Malta and Vibert [1997c]), as shown in the last section.

In what follows, we present some discussions about comparison of basin boundaries. One method to compare the boundaries separating the basins of attraction of the
equilibria r_1 and r_3 of (5.7.1) for different values of τ_1 and τ_2 is to consider a restricted
set of initial conditions such that there exists a one-to-one correspondence between
initial conditions for different values of τ_1 and τ_2. One method for doing this is to
consider constant functions as initial conditions. In the following, we will restrict our
attention to this class.

For a given real number c_1, let $\beta(c_1)$ be the unique real number such that $(c_1, \beta(c_1))$
is on the basin boundary. From the characterization of the basin boundary described
above, we know that the function β is a decreasing continuous function defined on the
real line. The graph of β divides the (c_1, c_2)-plane into two regions. Points "below"
this graph correspond exactly to the constant initial conditions lying in the basin of
attraction of r_1, and those "above" the graph to the ones belonging to the basin of
attraction of r_3. Thus B_c, the graph of β in the (c_1, c_2)-plane, can be considered as
the boundary separating the basins of attraction of the two stable equilibrium points.

To illustrate the effect of the delays on B_c, we consider identical neurons connected
with either symmetrical weights or symmetrical delays. In both cases, we set $I_1 = -\frac{w_1}{2}$ and $I_2 = \frac{w_2}{2}$. For these parameters, the coordinates of the equilibria r_1, r_2 and
r_3 are

$$r_1 = (-a, -b), \quad r_2 = (x_{1u}, x_{2u}) = (0, 0), \quad r_3 = (a, b),$$

where (a, b) is the strictly positive solution of (5.7.3). Note that in this case, the
system is invariant under the change $(x_1, x_2) \to (-x_1, -x_2)$. This implies that B_c is
also symmetric under the same transformation.

Symmetrical weights. For identical neurons, receiving the same inputs and connected
through symmetrical weights and delays, the neurons are indistinguishable from each
other. Therefore, the basin boundary B_c is the straight line defined by

$$c_1 - x_{1u} + c_2 - x_{2u} = 0.$$

Hence, in this situation, varying both delays together does not bring any change to
the basin boundary. However, when the delays are no longer equal, B_c undergoes
changes depending on the values of the delays. Numerical simulations in Pakdaman,
Grotta-Ragazzo, Malta, Arino and Vibert [1998b] show that when τ_2 decreases from
τ_1 to 0, the slope of the boundary at each point increases from -1 to a strictly negative
limit value which depends on τ_1.

Symmetrical delays. For neurons connected through symmetrical delays but non-
symmetrical weights, changes in the delays may modify the graph B_c. Again, numerical simulations in Pakdaman, Grotta-Ragazzo, Malta, Arino and Vibert [1998b] show
that the boundary B_c changes for nonsymmetric weights, when the delay is changed.
In particular, the slope of the boundary decreases as the delays $\tau_1 = \tau_2$ are increased
from 0 to 2, except at the unstable point r_2.

The two-neuron network studied here is representative for irreducible cooperative networks, since the theoretical description of the basin boundary, and more generally, the global picture of the phase portrait derived for the two-neuron network can be extended to irreducible cooperative networks of arbitrary size (Pakdaman, Malta, Grotta-Ragazzo and Vibert [1996]).

5.8   Synchronized activities

Assuming the network considered in (5.7.1) consists of two identical neurons without external inputs, we obtain the following:

$$\begin{cases} \dot{x}_1(t) = -\mu x_1(t) + f(x_2(t-\tau)), \\ \dot{x}_2(t) = -\mu x_2(t) + f(x_1(t-\tau)). \end{cases} \qquad (5.8.1)$$

In what follows, we will assume that $f : \mathbb{R} \to \mathbb{R}$ is C^2-smooth with $f(0) = 0$, $\lim_{x \to \pm\infty} f(x) = f_{\pm\infty} \in \mathbb{R}$, $f'(0) > \mu$ and $f'(x) > 0$ for all $x \in \mathbb{R}$, and $f''(x)x < 0$ for $x \neq 0$. It is easy to see that (5.8.1) has three equilibria (ξ_-, ξ_-), $(0, 0)$ and (ξ_+, ξ_+) with ξ^- and ξ^+ denoting the negative and positive zeros of

$$\mu\xi = f(\xi), \qquad (5.8.2)$$

and $\xi_\pm$ denoting the constant mapping on $[-1, 0]$ with the constant value $\xi^\pm$. Note also that $f'(\xi^\pm) < \mu$. See Figure 5.8.1.

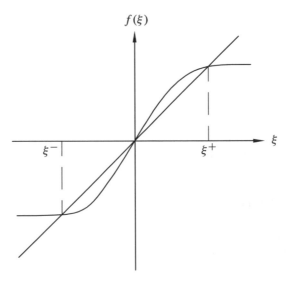

Figure 5.8.1. The graphs of $y = f(\xi)$ and $y = \xi$ intersect at (ξ^-, ξ^-), $(0, 0)$ and (ξ^+, ξ^+) with $f'(\xi^\pm) < \mu$.

Then results in Section 5.7 shows that almost every trajectory is convergent to (ξ_-, ξ_-) or (ξ_+, ξ_+). More precisely, there exists a closed subspace H of codimension 1 in the phase space $C([-\tau, 0]; \mathbb{R}^2)$ and a homeomorphism $g : H \to g(H) = B \subseteq C([-\tau, 0]; \mathbb{R}^2)$ such that the graph $B = g(H)$ is exactly the boundary separating the basins of attraction of (ξ_-, ξ_-) and (ξ_+, ξ_+). Moreover, this boundary consists of the initial value $(\varphi_1, \varphi_2) \in C([-\tau, 0]; \mathbb{R}^2)$ whose trajectories converge either to $(0, 0)$ or to a periodic orbit. In this section, we describe the detailed dynamics of the semiflow of (5.8.1) restricted to the boundary B and depict the geometric and topological structures of B.

Note also that the above discussions show that, due to the nature of identical decay rate μ and synaptic delay τ of the two neurons, almost every trajectory is eventually synchronized (convergent to the synchronous equilibria (ξ_-, ξ_-) or (ξ_+, ξ_+)), here a solution (x_1, x_2) of (5.8.1) is *synchronous* if $x_1 = x_2$ on the common domains of definition. Clearly, if we let z denote the common value of $x_1 = x_2$, then a synchronous solution is completely characterized by the scalar delay differential equation

$$\dot{z}(t) = -\mu z(t) + f(z(t - \tau)). \tag{5.8.3}$$

If we further normalize the delay τ by $x(t) = z(\tau t)$ then we get an equivalent system

$$\dot{x}(t) = -\tau \mu x(t) + \tau f(x(t - 1)).$$

Renaming the parameter and the signal function, we arrive at

$$\dot{x}(t) = -\mu x(t) + f(x(t - 1)) \tag{5.8.4}$$

with parameter $\mu > 0$ and a C^2-smooth bounded nonlinearity $f : \mathbb{R} \to \mathbb{R}$ satisfying all conditions stated in the above discussion. We now summarize the results from the monograph of Krisztin, Walther and Wu [1999] and explain the major steps towards these results.

Every element $\phi \in C([-1, 0]; \mathbb{R})$ determines a solution $x^\phi : [-1, \infty) \to \mathbb{R}$ of equation (5.8.4). The relations

$$F(t, \phi) = x_t^\phi$$

define a continuous semiflow $F : \mathbb{R}^+ \times C \to C$. Using the associated variation-of-constants formula

$$x(t) = e^{-\mu t} x(0) + \int_0^t e^{-\mu(t-s)} f(x(s - 1)) \, ds,$$

one can easily show that all maps $F(t, \cdot)$ with $t \geq 0$ are injective and continuously differentiable and F is monotone with respect to the pointwise ordering $<$ on C. The positively invariant set

$$S = \{\phi \in C; (x^\phi)^{-1}(0) \text{ is unbounded}\}$$

separates the domain of absorption into the interior of the positively invariant cone

$$K = \{\phi \in C; \phi(s) \geq 0 \text{ for all } s \in [-1, 0]\}$$

from the domain of absorption into the interior of $-K$. Such a separatrix has the following nonordering property: if φ, $\psi \in C$ and $\varphi < \psi$ then either $\varphi \in C \setminus S$ or $\psi \in C \setminus S$. Or equivalently, S does not contain two distinct elements which are ordered.

In order to derive a graph representation of S, we linearize equation (5.8.4) at the equilibrium 0. The derivatives $\{D_2 F(t, 0)\}_{t \geq 0}$ form a strongly continuous semigroup and satisfy

$$D_2 F(t, 0)\phi = y_t^\phi,$$

with the solution $y^\phi : [-1, \infty) \to \mathbb{R}$ of the linearized equation

$$\dot{y}(t) = -\mu y(t) + f'(0)y(t-1) \tag{5.8.5}$$

given by $y_0^\phi = \phi$. As discussed in Section 5.4, the spectrum σ of the generator of the semigroup consists of simple eigenvalues which coincide with the zeros of the characteristic function

$$\mathbb{C} \ni \lambda \mapsto \lambda + \mu - f'(0)e^{-\lambda} \in \mathbb{C}.$$

There is one real eigenvalue λ_0. The other eigenvalues form a sequence of complex conjugate pairs $(\lambda_j, \overline{\lambda_j})$ with

$$\operatorname{Re} \lambda_{j+1} < \operatorname{Re} \lambda_j < \lambda_0 \quad \text{and} \quad (2j-1)\pi < \operatorname{Im} \lambda_j < 2j\pi$$

for all integers $j \geq 1$, and $\operatorname{Re} \lambda_j \to -\infty$ as $j \to \infty$. The number of eigenvalues in the open right halfplane depends on μ and $f'(0)$. Any odd integer $2j + 1$, $j \in \mathbb{N}$, can be achieved. Let P, L, and Q denote the realified generalized eigenspaces of the generator associated with the spectral sets $\{\lambda_0\}$, $\{\lambda_1, \overline{\lambda_1}\}$, and $\sigma \setminus \{\lambda_0, \lambda_1, \overline{\lambda_1}\}$, respectively. Then

$$C = Q \oplus L \oplus P.$$

It is easy to see that $P = \mathbb{R}\chi_0$ with $\chi_0(\theta) = e^{\lambda_0 \theta}$ for $\theta \in [-1, 0]$.

We assume that

$$f'(0) > \mu/\cos\theta_\mu, \quad \text{where } \theta_\mu \in \left(\tfrac{3\pi}{2}, 2\pi\right) \text{ and } \theta_\mu = -\mu \tan\theta_\mu. \tag{5.8.6}$$

As shown in Section 5.4, this is equivalent to

$$0 < \operatorname{Re} \lambda_1. \tag{5.8.7}$$

Then the unstable space of the generator of the semigroup $\{D_2 F(t, 0)\}_{t \geq 0}$ is at least 3-dimensional.

In Krisztin, Walther and Wu [1999], we proved the following graph representation theorem.

Theorem 5.8.1. *There exists a map* $\mathrm{Sep} : Q \oplus L \to P$ *such that* $S = \{\chi + \mathrm{Sep}(\chi);$ $\chi \in Q \oplus L\}$ *and*

$$\| \mathrm{Sep}(\chi) - \mathrm{Sep}(\tilde{\chi}) \| \le e^{\lambda_0} \|\chi - \tilde{\chi}\| \quad \text{for } \chi, \tilde{\chi} \in Q \oplus L.$$

For each $\varphi \in C$, let φ^P, φ^Q, φ^L denote the projections of φ onto P, Q, L, respectively. If $\varphi \in C \setminus S$, then $\varphi^P \ne \mathrm{Sep}(\varphi^Q + \varphi^L)$. Since $P = \mathbb{R}\chi_0$ with $\chi_0 > 0$, we obtain

$$\varphi^P < \mathrm{Sep}(\varphi^Q + \varphi^L) \quad \text{or} \quad \varphi^P > \mathrm{Sep}(\varphi^Q + \varphi^L)$$

and we say in the first case, φ is below S and in the second case φ is above S.

It is easy to show that S is the boundary separating the basins of attraction of (ξ_-, ξ_-) and (ξ_+, ξ_+) in the following sense:

Theorem 5.8.2. *If* φ *is above* S *then* $\varphi_t^\varphi \to (\xi_+, \xi_+)$ *as* $t \to \infty$, *and if* φ *is below* S *then* $x_t^\varphi \to (\xi_-, \xi_-)$ *as* $t \to \infty$.

A local approximation of S near zero is possible by the linearization (5.8.5). To be more precise, we recall that there exists a 3-dimensional C^1-submanifold W_{loc} of C which is tangent at 0 to the realified eigenspace $L \oplus P$ of the spectral set $\{\lambda_0, \lambda_1, \overline{\lambda_1}\}$ and has the property that for every $\phi \in W_{\mathrm{loc}}$ there exist a solution $x : \mathbb{R} \to \mathbb{R}$ and $T \in \mathbb{R}$ such that $x_0 = \phi$ and $x_s \in W_{\mathrm{loc}}$ for all $s \le T$. The forward extension

$$W = F(\mathbb{R}^+ \times W_{\mathrm{loc}})$$

of W_{loc} is an invariant subset of C.

In the following, for a given Banach space E, we use E_ε to denote the open ball in E centered at the zero and of radius $\varepsilon > 0$. Then we have the following local approximation theorem for $W_{\mathrm{loc}} \cap S$.

Theorem 5.8.3. *There exist* $\eta > 0$ *and* $\varepsilon > 0$ *and a Lipschitz continuous map* $\mathrm{Sep}_\eta :$ $L_\eta \to Q \oplus P$ *such that*

$$\mathrm{Sep}_\eta(0) = 0, \quad \mathrm{Sep}_\eta(L_\eta) \subseteq Q \oplus P_\varepsilon,$$

$$(W_{\mathrm{loc}} \cap S) \cap (Q + L_\eta + P_\varepsilon) = \{\chi + \mathrm{Sep}_\eta(\chi); \chi \in L_\eta\}.$$

We refer to Figure 5.8.2 for illustration.

Recall that $W \subseteq C$ is invariant. Later we will see there exists a close relation between W and the global attractor of (5.8.4), and thus it is important to describe the structure of $\overline{W}$.

The following is the main result in Krisztin, Walther and Wu [1999].

Theorem 5.8.4. $\overline{W}$ *is invariant, and the semiflow defines a continuous flow* $F_W :$ $\mathbb{R} \times \overline{W} \to \overline{W}$. *For* $\overline{W}$ *and for the part* $\overline{W} \cap S$ *of the separatrix* S *in* $\overline{W}$, *we have the following graph representations (see Figure 5.8.3 for illustration):*

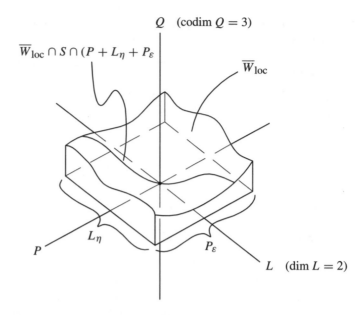

Figure 5.8.2. The local approximation of $W_{\mathrm{loc}} \cap S$.

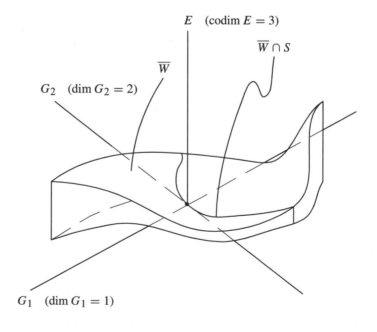

Figure 5.8.3. Graph representations of $\overline{W}$ and $\overline{W} \cap S$.

There exist subspaces $G_2 \subset G_3$ of C of dimensions 2 and 3, respectively, a complementary space G_1 of G_2 in G_3, a closed complementary space E of G_3 in C, a compact set $D_S \subset G_2$ and a compact set $D_W \subset G_3$, and continuous mappings $w : D_W \to E$ and $w_S : D_S \to G_1 \oplus E$ such that[9]

$$\overline{W} = \{\chi + w(\chi); \chi \in D_W\}, \quad \overline{W} \cap S = \{\chi + w_S(\chi); \chi \in D_S\}.$$

We have

$$D_W = \partial D_W \cup \overset{\circ}{D}_W, \quad W = \{\chi + w(\chi); \chi \in \overset{\circ}{D}_W\},$$

and the restriction of w to $\overset{\circ}{D}_W$ is C^1-smooth. The restriction of F_W to $\mathbb{R} \times W$ is C^1-smooth. The domain D_S is homeomorphic to the closed unit disk in $\mathbb{R}^2$, and consists of the trace of a simple closed C^1-curve and its interior. The map w_S is C^1-smooth in the sense that $w_S |_{\overset{\circ}{D}_S}$ is C^1-smooth, and for each $\chi \in \partial D_S$ there is an open neighborhood N of χ in G_2 so that $w_S |_{N \cap D_S}$ extends to a C^1-map on N.

Concerning the dynamics, we have that

the set

$$\mathcal{O} = \overline{W} \cap S \setminus (W \cap S) = \{\chi + w_S(\chi); \chi \in \partial D_S\}$$

is a periodic orbit, and there is no other periodic orbit in $\overline{W}$. The open annulus $(W \cap S) \setminus \{0\}$ consists of heteroclinic connections from 0 to $\mathcal{O}$. (Namely, $F(t, \phi) \to 0$ as $t \to -\infty$ and $F(t, \phi) \to \mathcal{O}$ as $t \to \infty$ for every ϕ in the open annulus). For every $\phi \in W$, $F_W(t, \phi) \to 0$ as $t \to -\infty$. $\overline{W}$ is compact and contains the equilibria (ξ_-, ξ_-) and (ξ_+, ξ_+) in C. For every $\phi \in \overline{W}$.

$$(\xi_-, \xi_-) \le \phi \le (\xi_+, \xi_+).$$

There exist homeomorphisms from $\overline{W}$ and D_W onto the closed unit ball in $\mathbb{R}^3$, which send

$$\text{bd } W = \overline{W} \setminus W = \{\chi + w(\chi); \chi \in \partial D_W\} \quad \text{and} \quad \partial D_W$$

onto the unit sphere $S^2 \subset \mathbb{R}^3$. If we define χ_- and χ_+ by $(\xi_-, \xi_-) = \chi_- + w(\chi_-)$ and $(\xi_+, \xi_+) = \chi_+ + w(\chi_+)$, respectively, then the set $\partial D_W \setminus \{\chi_-, \chi_+\}$ is a 2-dimensional C^1-submanifold of G_3, and the restriction of w to $D_W \setminus \{\chi_-, \chi_+\}$ is C^1-smooth (in the sense explained

---

[9]For a given set M of a topological space, we use $\overset{\circ}{M}$, $\overline{W}$ and ∂M to denote the interior, closure and boundary of M, respectively.

above for w_S). The points $\phi \in \overline{W} \setminus (\overline{W} \cap S)$ above the separatrix S form a connected set and satisfy $F_W(t, \phi) \rightarrow (\xi_+, \xi_+)$ as $t \rightarrow \infty$, and all $\phi \in \overline{W} \setminus (\overline{W} \cap S)$ below the separatrix S form a connected set and satisfy $F_W(t, \phi) \rightarrow (\xi_-, \xi_-)$ as $t \rightarrow \infty$. Finally, for every ϕ in the set bd W and different from (ξ_-, ξ_-) and (ξ_+, ξ_+), $F_W(t, \phi) \rightarrow \mathcal{O}$ as $t \rightarrow -\infty$.

Combining all the results stated so far, one may visualize the invariant set $\overline{W}$ as a smooth solid spindle which is split by an invariant disk into the basins of attraction towards the tips (ξ_-, ξ_-) and (ξ_+, ξ_+), see Figure 5.8.4.

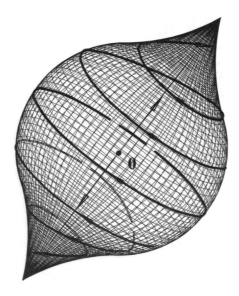

Figure 5.8.4. A smooth solid spindle in the attractor of a scalar delay positive feedback equation.

The first steps toward these results exploit the monotonicity of the semiflow. Also important is the observation that the set $\overline{W}$ is contained in the order interval between the stationary points (ξ_-, ξ_-) and (ξ_+, ξ_+), and that there are heteroclinic connections from 0 to (ξ_-, ξ_-) and (ξ_+, ξ_+), given by monotone solutions $x : \mathbb{R} \rightarrow \mathbb{R}$ without zeros.

An important tool for the investigation of finer structures is a version of the discrete Liapunov functional $V : C \setminus \{0\} \rightarrow \mathbb{N}$ counting sign changes, which was introduced by Mallet-Paret [1988]. Related are *a-priori* estimates for the growth and decay of solutions with segments in (sub-)level sets of V, which go back to Mallet-Paret [1988] and in a special case to Walther [1977, 1979]. These tools are first used to characterize the invariant sets $W \setminus \{0\}$ and $W \cap S \setminus \{0\}$ as the sets of segments x_t of solutions $x : \mathbb{R} \rightarrow \mathbb{R}$ with α-limit set $\{0\}$ which satisfy $V(x_t) \leq 2$ for all $t \in \mathbb{R}$ and $V(x_t) = 2$ for all $t \in \mathbb{R}$, respectively. Moreover, nontrivial differences of segments in $\overline{W}$ and

$\overline{W} \cap S$ belong to $V^{-1}(\{0, 2\})$ and $V^{-1}(2)$, respectively. The last facts permit to introduce global coordinates on $\overline{W}$ and $\overline{W} \cap S$. It is not difficult to show that the continuous linear evaluation map

$$\Pi_2 : C \ni \phi \mapsto (\phi(0), \phi(-1))^T \in \mathbb{R}^2$$

is injective on $\overline{W} \cap S$, and the continuous linear evaluation map $\Pi_3 : C \to \mathbb{R}^3$ given by

$$\Pi_3 \phi = (\phi(0), \phi(-1), c_P(\phi))^T$$

and

$$(\mathrm{Pr}_P \phi)(t) = \frac{1}{1 + f'(0)e^{-\lambda_0}} c_P(\phi) e^{\lambda_0 t},$$

where Pr_P is the spectral projection onto P along $Q \oplus L$, is injective on $\overline{W}$. The inverse maps of the restrictions of Π_2 and Π_3 to $\overline{W} \cap S$ and $\overline{W}$, respectively, turn out to be locally Lipschitz continuous.

The next step leads to the desired graph representation. We embed $\mathbb{R}^3 \supset \Pi_3 \overline{W}$ and $\mathbb{R}^2 \supset \Pi_2(\overline{W} \cap S)$ in a simple way as subspaces $G_3 \supset G_2$ into C, so that representations by maps w and w_S with domains in G_3 and G_2 and ranges in complements E of G_3 in C and $E \oplus G_1$ of G_2 in C, $G_1 \subset G_3$, become obvious. It is not hard to deduce that W is given by the restriction of w to an open set, and that this restriction is C^1-smooth. On $\overline{W}$, the semiflow extends to a flow $F_W : \mathbb{R} \times \overline{W} \to \overline{W}$, and F_W is C^1-smooth on the C^1-manifold $\mathbb{R} \times W$.

Phase plane techniques apply to the coordinate curves in $\mathbb{R}^2$ (or G_2) of the flowlines $t \mapsto F_W(t, \phi)$ in the invariant set $\overline{W} \cap S$, and yield the periodic orbit

$$\mathcal{O} = (\overline{W} \cap S) \setminus (W \cap S)$$

as well as the identification

$$\Pi_2(W \cap S) = \mathrm{int}(\Pi_2 \mathcal{O}),$$

which implies that also $W \cap S$ is given by the restriction of w_S to an open subset of G_2.

The investigation of the smoothness of the part $W \cap S$ of the separatrix S in W and of the manifold boundary

$$\mathrm{bd}\, W = \overline{W} \setminus W$$

starts with a study of the stability of the periodic orbit $\mathcal{O}$. We use the fact that there is a heteroclinic flowline in $W \cap S$ from the stationary point $0 \in C$ to the orbit $\mathcal{O}$, i.e., in the level set $V^{-1}(2)$, in order to show that precisely one Floquet multiplier lies outside the unit circle. It also follows that the center space of the linearized period map, or monodromy operator

$$M = D_2 F(\omega, p_0),$$

given by $p_0 \in \mathcal{O}$ and the minimal period $\omega > 0$, is at most 2-dimensional. The study of the linearized stability is closely related to earlier work in Lani-Wayda and Walther [1995] and to *a-priori* results on Floquet multipliers and eigenspaces for general monotone cyclic feedback systems with delay due to Mallet-Paret and Sell [1996a].

To show that the graph $W \cap S \subset V^{-1}(2) \cup \{0\}$ is C^1-smooth. We identify pieces of $W \cap S$ in a hyperplane Y transversal to $\mathcal{O}$, as open sets in the smooth transversal intersection of W with the center-stable manifold of the Poincaré return map which is associated with Y and a point $p_0 \in \mathcal{O}$. Then we use the C^1-flow F_W to obtain the smoothness of the set $W \cap S \setminus \{0\}$. (Smoothness close to 0 and the relation

$$T_0(W \cap S) = L$$

follow by other arguments.) Of course, this approach relies on a result on the existence of C^1-smooth center-stable manifolds at fixed points of C^1-maps in Banach spaces.

A technical aspect of the smoothness proof for $W \cap S$ is that close to a fixed point of a C^1-map with one-dimensional linear center space we have to construct open and positively invariant subsets of the center-stable manifold W^{cs} provided a single trajectory in W^{cs} but not in the strong stable manifold converges to the fixed point.

Having established the smoothness of $W \cap S$ we show in a series of propositions that the manifold boundary bd $W = \overline{W} \setminus W$ (without the stationary points (ξ_-, ξ_-) and (ξ_+, ξ_+)) coincides with the forward extensions of a local unstable manifold of the period map $F(\omega, \cdot)$ at a fixed point $p_0 \in \mathcal{O}$. The long proof of this fact involves the charts Π_2 and Π_3, and uses most of the results obtained before.

The next step achieves the smoothness of a piece of $\overline{W}$ in a hyperplane H of C transversal to the periodic orbit $\mathcal{O}$. We construct a C^1-smooth graph over an open set in a plane $X_{12} \subset H$ which extends such a piece of $\overline{W} \cap H$ close to a point $p_0 \in \mathcal{O} \cap H$ beyond the boundary. Using the flow F_W we then derive that $\overline{W}$ and $\overline{W} \cap S$ are C^1-smooth up to their manifold boundaries, in the sense stated before.

The final steps lead to the topological description of $\overline{W}$. First we use the characterization of bd W mentioned above to define homeomorphisms from bd W onto the unit sphere $S^2 \subset \mathbb{R}^3$. Then a generalization due to Bing [1983] of the Schoenfliess theorem from planar topology is employed to obtain homeomorphisms from $\overline{W}$ onto the closed unit ball in $\mathbb{R}^3$. The application of Bing's theorem requires to identify the bounded component of the complement of the set

$$\Pi_3(\text{bd } W) \cong S^2$$

in $\mathbb{R}^3$ as the set $\Pi_3 W$, and to verify that $\Pi_3 W$ is uniformly locally 1-connected. This means that for every $\varepsilon > 0$ there exists $\delta > 0$ so that every closed curve in a subset of $\Pi_3 W$ with diameter less than δ can be continuously deformed to a point in a subset of $\Pi_3 W$ with diameter less than ε. We point that the proof of this topological property relies on the smoothness of the set

$$\overline{W} \setminus \{(\xi_-, \xi_-), (\xi_+, \xi_+)\}$$

and involves subsets of boundaries of neighborhoods of 0, (ξ_-, ξ_-), (ξ_+, ξ_+) in W which are transversal to the flow F_W. In order to construct these smooth boundaries we have to go back to the variation-of-constants formula for retarded functional differential equations in the framework of sun-dual and sun-star dual semigroups (Diekmann, van Gils, Verduyn Lunel and Walther [1995]).

We note that for some general sigmoid functions (including $f(\xi) = \tanh(\alpha\xi)$ with some $\alpha > 0$) under the condition equivalent to Re $\lambda_2 < 0$, Krisztin and Walther [1999] proved that $\overline{W}$ is indeed the global attractor of the semiflow generated by (5.8.4). Note also that the solid spindle $\overline{W}$ loses its smoothness at the two tips (ξ_-, ξ_-) and (ξ_+, ξ_+), the singularity at these two points have been completely characterized by Walther [1998].

As a final remark, we note that system (5.8.3) can be rewritten as

$$\varepsilon \dot{w}(t) = -\mu w(t) + f(w(t-1))$$

with $\varepsilon = \tau^{-1}$. Therefore, the case of large delay corresponds to the situation where $\varepsilon \to 0$. This motivates the study of Chen [1999] on the limiting profiles of the periodic solutions for the equation

$$\varepsilon \dot{x}(t) = -x(t) + f(x(t-1), \lambda) \tag{5.8.8}$$

with $f \in C^3(\mathbb{R} \times \mathbb{R}; \mathbb{R})$, $f(0, \lambda) = 0$ and

$$f(x, \lambda) = (1 + \lambda)x + ax^2 + bx^3 + o(x^3) \text{ as } x \to 0.^{10}$$

It was shown that in case where $a = 0$, there is a neighborhood U of $(0, 0)$ in the (λ, ε) plane and a sectorial region S in U such that if $(\lambda, \varepsilon) \in U$, then there is a unique periodic solution $x_{\lambda,\varepsilon}$ of (5.8.8) with period $1 + \varepsilon + O(|\varepsilon|(|\lambda| + |\varepsilon|))$ as $(\lambda, \varepsilon) \to (0, 0)$ if and only if $(\lambda, \varepsilon) \in S$. Moreover, if $b < 0$ then for small and fixed $\lambda = \lambda_0 > 0$, the set $\{\varepsilon; (\lambda_0, \varepsilon) \in S\}$ is an interval $(0, \varepsilon(\lambda_0))$. At the point $(\lambda_0, \varepsilon(\lambda_0))$, there is a Hopf bifurcation of sine waves and the periodic solution approaches a *square wave* as $\varepsilon \to 0$. That is, the periodic solution $x_{\lambda_0,\varepsilon}(t) \to c_{1\lambda_0}$ (respectively, $c_{2\lambda_0}$) as $\varepsilon \to 0$ uniformly on compact subsets of $(0, \frac{1}{2})$ (respectively, $(\frac{1}{2}, 1)$), where $c_{1\lambda_0}$ and $c_{2\lambda_0}$ are the two distinct nontrivial fixed points of $f(\cdot, \lambda_0)$. If $b > 0$ then for small and fixed $\lambda = \lambda_0 > 0$, the set $\{\varepsilon; (\lambda_0, \varepsilon) \in S\}$ is an interval $(\varepsilon(\lambda_0), \beta(\lambda_0))$. At the point $(\lambda_0, \varepsilon(\lambda_0))$, there is a Hopf bifurcation. For small and fixed $\lambda = \lambda_0 < 0$, the set $\{\varepsilon; (\lambda_0, \varepsilon) \in S\}$ is an interval $(0, \alpha(\lambda_0))$. As $\varepsilon \to 0$, the unique solution becomes *pulse*-like in the following sense: the periodic solution $x_{\lambda_0,\varepsilon}$ has the property that $x_{\lambda_0,\varepsilon}(t) \to 0$ as $\varepsilon \to 0$ uniformly on compact subsets of $(0, \frac{1}{2}) \cup (\frac{1}{2}, 1)$. The magnitude of the pulse exceeds $\max\{|c_{1\lambda_0}|, |c_{2\lambda_0}|\}$. See Figures 5.8.5 and 5.8.6.

---

[10]The case where

$$f(x, \lambda) = -(1 + \lambda)x + ax^2 + bx^3 + o(x^3) \text{ as } x \to 0$$

was discussed by Chow, Hale and Huang [1992] and Hale and Huang [1994].

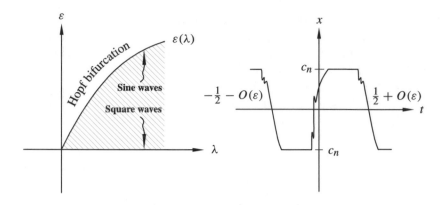

Figure 5.8.5. Limiting profiles of periodic solutions when $a = 0$ and $b < 0$: square waves.

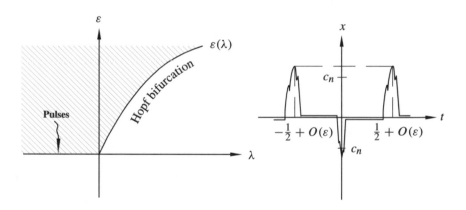

Figure 5.8.6. Limiting profiles of periodic solutions when $a = 0$ and $b > 0$: pulses.

In the case where $a \neq 0$, there exist a neighborhood U of $(0, 0)$ and a sectorial region S in U such that if $(\lambda, \varepsilon) \in U$, then there is a unique periodic orbit, determined by the solution $x_{\lambda,\varepsilon}$ of (5.8.8) with period $1 + \varepsilon + O(|\varepsilon|(|\lambda| + |\varepsilon|))$ as $(\lambda, \varepsilon) \to (0, 0)$ if and only if $(\lambda, \varepsilon) \in S$. Moreover, for small and fixed $\lambda = \lambda_0 > 0$, the set $\{\varepsilon; (\lambda_0, \varepsilon) \in S\}$ is an interval $(\varepsilon(\lambda_0), \beta(\lambda_0))$. At the point $(\lambda_0, \varepsilon(\lambda_0))$, there is a Hopf bifurcation. For small and fixed $\lambda = \lambda_0 < 0$, the set $\{\varepsilon; (\lambda_0, \varepsilon) \in S\}$ is an interval $(0, \alpha(\lambda_0))$. As $\varepsilon \to 0$, the unique periodic solution becomes pulse-like with the magnitude of the pulse exceeding $|c_{0\lambda_0}|$, where $c_{0\lambda_0}$ is the nontrivial fixed point of $f(\cdot, \lambda_0)$ in a small neighborhood of 0. See Figure 5.8.7 for the limiting profiles in such a case.

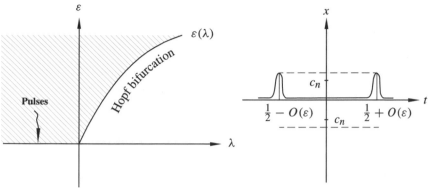

Figure 5.8.7. Limiting profiles for the case where $a \neq 0$.

5.9 Desynchronization, phase-locked oscillation and connecting orbits

We continue our discussion about the following system of delay differential equations

$$\begin{cases} \dot{u}(t) = -\mu u(t) + f(v(t - \tau)), \\ \dot{v}(t) = -\mu v(t) + f(u(t - \tau)), \end{cases}$$

or equivalently (after rescaling the time variable)

$$\begin{cases} \dot{u}(t) = -\mu \tau u(t) + \tau f(v(t - 1)), \\ \dot{v}(t) = -\mu \tau v(t) + \tau f(u(t - 1)) \end{cases} \qquad (5.9.1)$$

for a network of two identical neurons, where τ and μ are given positive constants and $f : \mathbb{R} \to \mathbb{R}$ is a C^1-map satisfying the following set of conditions:

$f(0) = 0$, $f'(\xi) > 0$ for all $\xi \in \mathbb{R}$ (Monotone Positive Feedback);

$f(\xi) = \mu \xi$ has exactly three zeros $\xi^- < 0 < \xi^+$ and $\max\{f'(\xi^-), f'(\xi^+)\} < \mu < f'(0)$ (Dissipativeness and Instability of 0);

$f(\xi) = -f(-\xi)$ for all $\xi \in \mathbb{R}$ (Symmetry);

The function $(0, \infty) \ni \xi \mapsto \frac{\xi f'(\xi)}{f(\xi)} \in \mathbb{R}$ is monotonically decreasing (Concavity).

It should be mentioned that some results described in this section (obtained in Chen, Krisztin and Wu [1999]) do not require the symmetry and concavity conditions on f, and that the function

$$f(\xi) = \arctan(\gamma \xi), \quad \xi \in \mathbb{R}$$

or

$$f(\xi) = \frac{e^{\gamma\xi} - e^{-\gamma\xi}}{e^{\gamma\xi} + e^{-\gamma\xi}}, \quad \xi \in \mathbb{R}$$

satisfies the above set of conditions with $\gamma > \mu$.

Our ultimate goal is to describe the global attractor of system (5.9.1), and our focus here is on the existence of connecting orbits from a synchronous periodic solution to a phase-locked periodic solution, here a *phase-locked* T-periodic solution of (5.9.1) is a solution satisfying $u(t) = v(t - \frac{T}{2})$ for all $t \in \mathbb{R}$. The existence of such connecting orbits suggests a mechanism for desynchronization of the network.

As shown in the last section, a synchronous solution of (5.9.1) is completely governed by the scalar delay differential equation for $w(= u = v)$:

$$\dot{w}(t) = -\mu\tau w(t) + \tau f(w(t-1)). \tag{5.9.2}$$

The zeros of the characteristic equation

$$\lambda + \mu\tau - \tau f'(0)e^{-\lambda} = 0 \tag{5.9.3}$$

form the spectrum of the generator A_s of the solution semiflow of the linear equation

$$\dot{Z}(t) = -\mu\tau Z(t) + \tau f'(0)Z(t-1). \tag{5.9.4}$$

All zeros of (5.9.3) are simple eigenvalues of A_s. There is one real eigenvalue, the other eigenvalues form complex conjugate pairs with real parts less than the value of the real eigenvalue. Defining

$$\tau_s = \frac{2\pi - \arccos\frac{\mu}{f'(0)}}{\sqrt{[f'(0)]^2 - \mu^2}}, \qquad \tau_s' = \frac{4\pi - \arccos\frac{\mu}{f'(0)}}{\sqrt{[f'(0)]^2 - \mu^2}},$$

one finds that for $\tau > \tau_s$ the realified generalized eigenspace of A_s associated with the spectral set of the eigenvalues with positive real part is at least 3-dimensional, while for $\tau_s < \tau < \tau_s'$ it is exactly 3-dimensional. When $\tau > \tau_s$, there exists a 3-dimensional local unstable manifold $\hat{W}_{3,s,\mathrm{loc}}$ of the solution semiflow generated by equation (5.9.2), which is, at zero, tangent to the 3-dimensional realified generalized eigenspace of the generator A_s associated with the positive real eigenvalue and the pair of complex conjugate eigenvalues with the greatest real part. We know that $\overline{\hat{W}_{3,s}}$ is a 3-dimensional submanifold of C which contains a smooth invariant disk bordered by a periodic orbit. The disk separates $\overline{\hat{W}_{3,s}}$ into halves: one half contains connecting orbits from the stationary point 0 or the periodic orbit to a positive stationary point, the other half contains connecting orbits from the stationary point 0 or the periodic orbit to a negative stationary point. Moreover, if $\tau \in (\tau_s, \tau_s')$ then $\overline{\hat{W}_{3,s}}$ is indeed the global attractor of the solution semiflow of equation (5.9.2).

It is interesting to note that the characteristic equation of the linearization at 0 of the full system (5.9.1), namely,

$$\begin{cases} \dot{X}(t) = -\mu\tau X(t) + \tau f'(0)Y(t-1), \\ \dot{Y}(t) = -\mu\tau Y(t) + \tau f'(0)X(t-1) \end{cases} \qquad (5.9.5)$$

can be decoupled as

$$[\lambda + \mu\tau - \tau f'(0)e^{-\lambda}][\lambda + \mu\tau + \tau f'(0)e^{-\lambda}] = 0, \qquad (5.9.6)$$

where the first factor corresponds to the characteristic equation (5.9.3). Defining

$$\tau_d = \frac{\pi - \arccos\frac{\mu}{f'(0)}}{\sqrt{[f'(0)]^2 - \mu^2}},$$

one sees that for $\tau > \tau_d$ the zeros of the equation $\lambda + \mu\tau + \tau f'(0)e^{-\lambda} = 0$ are simple and occur in complex conjugate pairs, at least one pair of complex conjugate zeros has positive real part, while for $0 < \tau \leq \tau_d$ all zeros have nonpositive real parts. In Chen and Wu [1998], it was proved that when $\tau > \tau_d$, there exists a complete analogue $\overline{\hat{W}}_{3,d}$ of $\overline{\hat{W}}_{3,s}$ for the full system (5.9.1). In particular, this $\overline{\hat{W}}_{3,d}$ is separated by a smooth invariant disk borded by a phase-locked periodic orbit $\hat{O}_2$.

Consequently, when $\tau > \tau_s (> \tau_d)$, the unstable space of the generator of the C_0-semigroup generated by (5.9.5) is at least 5-dimensional and there exists a 5-dimensional local unstable manifold $\hat{W}_{5,\text{loc}}$ tangent, at zero, to the realified generalized eigenspace of the generator associated with the positive real and the two leading pairs of complex conjugate eigenvalues with positive real parts. The global forward extension $\hat{W}_5$ of $\hat{W}_{5,\text{loc}}$ contains $\overline{\hat{W}}_{3,d}$ and $\widetilde{\hat{W}}_{3,s} = \{(\varphi, \varphi); \varphi \in \hat{W}_{3,s}\}$, and in particular, $\overline{\hat{W}}_5$ contains a phase-locked periodic orbit $\hat{O}_2$ and a synchronous periodic orbit $\hat{O}_4$. Chen, Krisztin and Wu [1999] described completely the global dynamics of the semiflow of system (5.9.1) restricted to $\overline{\hat{W}}_5$ and established the existence of heteroclinic orbits from the synchronous periodic orbit $\hat{O}_4$ to the phase-locked periodic orbit $\hat{O}_2$.

The purpose of this section is to briefly describe the main results, technical tools and approach of the above work. First of all, we make the following change of variables

$$\begin{cases} x(t) = u(2t), \\ y(t) = v(2t-1) \end{cases}$$

to get an equivalent cyclic system of delay differential equations

$$\begin{cases} \dot{x}(t) = -2\mu\tau x(t) + 2\tau f(y(t)), \\ \dot{y}(t) = -2\mu\tau y(t) + 2\tau f(x(t-1)). \end{cases} \qquad (5.9.7)$$

Powerful technical tools and general results have been developed in the recent work of Mallet-Paret and Sell [1996a, b].

In what follows, we are going to describe the results only for the transformed system (5.9.7), with O_4, O_2, $W_{3,d}$, W_5 and $W_{3,s}$ denoting the analogues of $\hat{O}_4$, $\hat{O}_2$, $\hat{W}_{3,d}$, $\hat{W}_5$ and $\tilde{W}_{3,s}$, respectively. The fundamental technical tool for (5.9.7) is an integer-valued Liapunov functional $V : C(\mathbb{K}) \setminus \{0\} \to \{0, 2, \ldots, \infty\}$, where $\mathbb{K} = [-1, 0] \cup \{1\}$, $C(\mathbb{K}) = \{\varphi : \mathbb{K} \to \mathbb{R}; \ \varphi \text{ is continuous }\}$ is the phase space of system (5.9.7) and $V(\varphi)$ measures, roughly, the number of sign changes of $\varphi \in C(\mathbb{K}) \setminus \{0\}$. Important properties of V have been established in Mallet-Paret and Sell [1996a], applications of these properties show that solutions of (5.9.7) can not decay too fast at ∞ (Theorem 4.1 of Chen and Wu [1998]) for (5.9.7) and lead to the following characterization of W_5:

$$(T1) \quad W_5 \setminus \{0\} = \left\{ \varphi \in C(\mathbb{K}) \setminus \{0\}; \quad \begin{array}{l} \text{there exists a solution } z^\varphi : \mathbb{R} \to \mathbb{R}^2 \\ \text{of (5.9.7) with } z_0^\varphi = \varphi, \ V(z_t^\varphi) \le 4 \\ \text{for all } t \in \mathbb{R} \text{ and } z_t^\varphi \to 0 \text{ as } t \to -\infty \end{array} \right\}.$$

Also

$$(T2) \qquad\qquad V(\varphi - \psi) \le 4 \quad \text{if } \varphi, \psi \in \overline{W}_5 \text{ and } \varphi \ne \psi.$$

Another important observation we make is that

$(T3)$ \qquad every non-constant periodic solution of (5.9.1) is either synchronous or phase locked.

This, together with the concavity and symmetry conditions, enables us to apply the results of Krisztin and Walther [1999] about the uniqueness of periodic solutions in each level set of V to show that

$(T4)$ \qquad there exists at most one periodic orbit in each level set $V^{-1}(2k)$, $k = 1, 2, \ldots$, of V.

One can also apply this Liapunov functional to show that

$(T5)$ \qquad there exists no homoclinic connections of O_2 and O_4.

The most important application of V is perhaps the proof of the Poincaré–Bendixson Theorem for cyclic systems (Mallet-Paret and Sell [1996b]) which, together with the existence of a heteroclinic connection from 0 to non-trivial equilibria of (5.9.7) (Hirsch [1988] or Smith [1987]), enables us to completely describe the dynamics of the semiflow on W_5. To be more precise, we introduce the separatrix

$$S = \{\varphi \in C(\mathbb{K}); \ \varphi = 0 \text{ or } V(z_t^\varphi) > 0 \text{ for all } t \ge 0\}.$$

Then we show that
$(T6)$

$$\begin{cases} \text{if } \varphi \in (W_5 \cap S) \setminus \{0\}, \text{ then } \alpha(\varphi) = \{0\}, \ \omega(\varphi) = O_2 \text{ or } O_4, \\[2mm] \text{if } \varphi \in W_5 \setminus S, \text{ then } \alpha(\varphi) = \{0\}, \ \omega(\varphi) = \{\chi_+\} \text{ or } \{\chi_-\}, \\[2mm] \text{if } \varphi \in \text{bd } W_5 \setminus (S \cup \{\chi_-, \chi_+\}), \text{ then } \alpha(\varphi) = O_4 \text{ or } O_2, \ \omega(\varphi) = \{\chi_+\} \text{ or } \{\chi_-\}, \\[2mm] \text{if } \varphi \in (\text{bd } W_5 \cap S) \setminus \{O_2 \cup O_4\}, \text{ then } \alpha(\varphi) = O_4 \text{ and } \omega(\varphi) = O_2, \end{cases}$$

where bd $W_5 = \overline{W}_5 \setminus W_5$.

To show that there is indeed a connecting orbit from O_4 to O_2, we need further information about the Floquet multipliers of O_4. We apply the general theory of Mallet-Paret and Sell [1996a] and some type of homotopy argument to show that

(T7)   there exist exactly 3 Floquet multipliers of O_4, counting multiplicities, outside the unit circle, and the associated realified generalized eigenspace is contained in $V^{-1}(\{0, 2\}) \cup \{0\}$.

An immediate consequence of (T7) implies that there exists a 3-dimensional C^1-smooth local unstable manifold $W^u_{\mathrm{loc}}(P)$ of a Poincaré-map P associated with a certain hyperplane. This manifold contains at least one point of a connecting orbit from O_4 to χ_+ and at least one point of a connecting orbit from O_4 to χ_-. Since $W^u_{\mathrm{loc}}(P)$ is 3-dimensional, there exists a continuous curve in $W^u_{\mathrm{loc}}(P) \setminus O_4$ connecting these points, and a continuity argument shows that the curve contains a point $\varphi \in W^u_{\mathrm{loc}}(P) \cap S$, which gives that

(T8)   there exists a connecting orbit from O_4 to O_2.

The above results can be depicted in Figure 5.9.1.

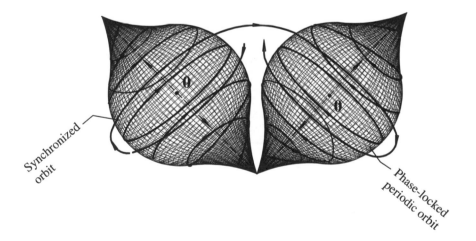

Figure 5.9.1. Two solid spindles and their connecting orbits in a system of two identical neurons. The left spindle is separated by a disk bordered by a phase-locked oscillation, and the right spindle is separated by a synchronous periodic orbit. There exist connecting orbits from the synchronized to the phase-locked orbit.

There are several questions remaining to be answered. Denoting by $W^u(O_4)$ the forward extension of $W^u_{\text{loc}}(P)$ we conjecture that

(C)
$$(\text{bd } W_5 \cap S) \setminus \{O_2 \cup O_4\} = (W^u(O_4) \cap S) \setminus \{O_2 \cup O_4\}$$
$$= \{\varphi \mid \alpha(\varphi) = O_4 \text{ and } \omega(\varphi) = O_2\}.$$

Also, note that
$$\varphi \in (\overline{W}_5 \cap S) \setminus \{0\} \Rightarrow \omega(\varphi) = O_2 \text{ or } O_4.$$

It is thus important to describe

$$H_i = \{\varphi \in (\overline{W}_5 \cap S) \setminus \{0\} \mid \omega(\varphi) = O_i\}, \quad i = 2, 4$$

(the basins of attraction of O_i in $\overline{W}_5$). Finally, we note that the next critical value of τ when (5.9.6) has a pair of purely imaginary zeros is $\tau_{2,d} = \dfrac{3\pi - \arccos \frac{\mu}{f'(0)}}{\sqrt{[f'(0)]^2 - \mu^2}} (> \tau_s)$. It is then reasonable to expect that $\overline{W}_5$ is the global attractor of (5.9.1) if $\tau \in (\tau_s, \tau_{2,d})$.

Limiting profiles of the synchronous periodic orbit and the phase-locked periodic solution have been described in Chen and Wu [1999c], pulses and square waves are both observed.

Bibliography

D. H. Ackley, G. H. Hinton and T. J. Sejnowski [1985], A learning algorithm for Boltzman machines. *Cognitive Science* **9**, 147–176.

U. an der Heiden [1980], *Analysis of Neural Networks*. Lecture Notes in Biomath. 35, Springer-Verlag, Berlin.

P. Andersen, G. N. Gross, T. Lomo and O. Sveen [1969], Participation of inhibitory and excitatory interneurons in the control of hippocampus cortical output. In: *The Interneuron* (M. Brazier, ed.). University of California Press, Los Angeles, 415–465

J. A. Anderson, J. W. Silverstein, S. A. Ritz and R. S. Jones [1977], Distinctive features, categorical perception, and probability learning: some applications of a neuron model. *Psychol. Rev.* **84**, 413–451.

K. L. Babcock and R. M. Westevelt [1987], Dynamics of simple electronic neural networks. *Phys. D* **28**, 305–316.

P. Baldi and A. F. Attyia [1994], How delays effect neural dynamics and learning. *IEEE Trans. Neural Networks* **5**, 612–621.

J. Bélair [1993], Stability in a model of a delayed neural network. *J. Dynam. Differential Equations* **5**, 607-623.

J. Bélair, S. A. Campbell and P. van den Driessche [1996], Frustration, stability and delay-induced oscillations in a neural network model. *SIAM J. Appl. Math.* **46**, 245–255.

J. Bélair and S. Dufour [1996], Stability in a three dimensional system of delay-differential equations. *Canad. Appl. Math. Quart.* **4**, 137–154.

A. Berman and R. J. Plemmons [1979], *Nonnegative Matrices in the Mathematical Sciences*. Academic Press, New York.

R. H. Bing [1983], *The Geometric Topology of 3-Manifolds*. Amer. Math. Soc. Colloq. Publ. 40. Providence, RI.

L. P. Burton and W. M. Whyburn [1952], Minimax solutions of ordinary differential systems. *Proc. Amer. Math. Soc.* **3**, 794–803.

S. Campbell and J. Bélair [1995], Analytical and symbolically-assisted investigation of Hopf bifurcations in delay-differential equations. *Canad. Appl. Math. Quart.* **3**, 137–154.

J. Cao and D. Zhou [1998], Stability analysis of delayed cellular neural networks. *Neural Networks* **11**, 1601–1605.

G. A. Carpenter [1994], A distributed outstar network for spatial pattern learning. *Neural Networks* **7**, 159–168.

Y. Chen [1999], Limiting profiles for periodic solutions of scalar delay differential equations. To appear in *Proc. Roy. Soc. Edinburgh*.

Y. Chen, T. Krisztin and J. Wu [1999], Connecting orbits from synchronous periodic solutions to phase-locked periodic solutions in a delay differential system. *J. Differential Equations* **163**, 130–173.

Y. Chen and J. Wu [1998], Existence and attraction of a phase-locked oscillation in a delayed network of two neurons. To appear in *Adv. Differential Equations*.

Y. Chen and J. Wu [1999a], Minimal instability and unstable set of a phase-locked periodic orbit in a delayed neural network. *Phys. D* **134**, 185–199.

Y. Chen and J. Wu [1999b], On a unstable set of a phase-locked periodic orbit of a network of two neurons. To appear in *J. Math. Anal. Appl.*

Y. Chen and J. Wu [1999c], The asymptotic shapes of perioic solutions of a singular delay differential system. To appear in *J. Differential Equations*.

S. N. Chow, J. K. Hale and W. Huang [1992], From sine waves to square waves in delay equations. *Proc. Roy. Soc. Edinburgh* **120A**, 223–229.

L. O. Chua and L. Yang [1988a], Cellular neural networks: theory. *IEEE Trans. Circuits Systems I* **35**, 1257–1272.

L. O. Chua and L. Yang [1988], Cellular neural networks: applications. *IEEE Trans. Circuits Systems I* **35**, 1273–1290.

P. S. Churchland and T. J. Sejnowski [1992], *The Computational Brain*. MIT Press, Cambridge.

M. A. Cohen [1990], The stability of sustained oscillation in symmetric coopertive-competitive networks. *Neural Networks* **3**, 609–612.

M. A. Cohen and S. Grossberg [1983], Absolute stability of global pattern formation and parallel memory storage by competitive neural networks. *IEEE Trans. Systems Man Cybernet.* **13**, 815–826.

F. H. C. Crick and C. Asanuma [1988], Certain aspects of the anatomy and physiology of cerebral cortex. In: *Parallel Distributed Processing* (J. L. McClell and D. E. Rumelhart, eds.). The MIT Press, Cambridge, Vol. 2, 333–371.

DARPA [1988], *DARPA Neural Network Study — Final Report* (Technical Report 840). Lincoln, MA: MIT Lincoln Laboratory.

O. Diekmann, S. A. van Gils, S. M. Verduyn Lunel and H.-O. Walther [1995], *Delay Equations: Functional-, Complex-, and Nonlinear Analysis*. Appl. Math. Sci. 110. Springer-Verlag, New York.

R. D. Driver [1977], *Ordinary and Delay Differential Equations*. Appl. Math. Sci. 20. Springer-Verlag, New York.

J. C. Eccles, M. Ito and J. Szentagothai [1967], *The Cerebellum as a Neuronal Machine*. Springer-Verlag, New York.

L. H. Erbe, K. Geba, W. Krawcewicz and J. Wu [1992], S^1-degree and global Hopf bifurcation theory of functional differential equations. *J. Differential Equations* **98**, 277–298.

T. Faria and L. T. Magalhães [1995], Normal forms for retarded functional differential equations with parameters and applications to Hopf bifurcations. *J. Differential Equations* **122**, 181–200.

B. Fiedler and T. Gedeon [1998], A class of convergence neural network dynamics. *Phys. D* **111**, 288–294.

J. M. Fuster, R. H. Bauer and J. P. Jerrey [1982], Cellular discharge in the dorsolateral prefrontal cortex of the monkey during cognitive tasks. *Exp. Neurol.* **77**, 679–694.

T. Gedeon [1999], Structure and dynamics of artificial neural networks. In: *Differential Equations with Applications to Biology* (S. Ruan, G. Wolkowicz and J. Wu, eds.). Fields Inst. Commun. 21. Amer. Math. Soc., Providence, RI, 217–224.

F. Giannakopoulos and A. Zapp [1999], Local and global Hopf bifurcation in a scalar delay differential equation, *Adv. Differential Equations* **237**, 425–450.

M. Gilli [1993], Strange attractors in delayed cellular neural networks, *IEEE Trans. Circuits Systems I* **40**, 849–853.

P. S. Goldman-Rakic [1984], Modular organization of prefrontal cortex, *Trends Neurosci.* **7**, 419–429.

K. Gopalsamy and X. Z. He [1994], Stability in asymmetric Hopfield nets with transmission delays, *Phys. D* **76**, 344–358.

S. Grossberg [1967], Nonlinear difference-differential equations in predictions and learning theory, *Proc. Nat. Acad. Sci. U.S.A.* **58**, 1329–1334.

S. Grossberg [1968a], Some physiological and biochemical consequences of psychological postulates, *Proc. Nat. Acad. Sci. U.S.A.* **60**, 758–765.

S. Grossberg [1968b], Global ratio limit theorems for some nonlinear functional-differential equations, I, II. *Bull. Amer. Math. Soc.* **74**, 95–99; 100–105.

S. Grossberg [1968c], A prediction theory for some nonlinear functional-differential equations, I. *J. Math. Anal. Appl.* **21**, 643–694.

S. Grossberg [1968d], A prediction theory for some nonlinear functional-differential equations, II. *J. Math. Anal. Appl.* **22**, 490–527.

S. Grossberg [1969], On learning and energy-entropy dependence in recurrent and nonrecurrent signed networks. *J. Statist. Physics* **1**, 319–350.

S. Grossberg [1970a], Some networks that can learn, remember, and reproduce any number of complicated patterns, II. *Stud. Appl. Math.* **49**, 135–166.

S. Grossberg [1970b], Embedding fields: underlying philosophy, matehmatics, and applications to psychology, physiology and anatomy. *J. Cybernet.* **1**, 28–50.

S. Grossberg [1970c], Neural pattern discrimination. *J. Theor. Biol.* **27**, 291–337.

S. Grossberg [1971a], Pavlovian pattern learning by nonlinear neural networks. *Proc. Nat. Acad. Sci. U.S.A.* **68**, 828–831.

S. Grossberg [1971b], On the dynamics of operating conditioning. *J. Theor. Biol.* **33**, 225–255.

S. Grossberg [1972], Pattern learning by functional-differential neural networks with arbitrary path weights. In: *Delay and Functional Differential Equations and Their Applications* (K. Schmitt, ed.), Academic Press, New York, 121–160.

S. Grossberg [1973], Contour enhancement, short term memory and constancies in reverberating neural networks. *Stud. Appl. Math.* **52**, 217–257.

S. Grossberg [1978a], A theory of human memory: self-organization and performance of sensory-motor codes, maps, and plans. In: Progr. Theoret. Biology (R. Rosen and F. Snell, eds.), Vol. 5, 233–274.

174   Bibliography

S. Grossberg [1978b], Competition, decision, and consensus. *J. Math. Anal. Appl.* **66**, 447–493.

S. Grossberg [1982], *Studies of Mind and Brain*. Boston Stud. Philos. Sci. 70. D. Reidel Publishing Co., Boston.

S. Grossberg [1988], Nonlinear networks: principles, mechanisms, and architectures. *Neural Networks* **1**, 17–61.

J. K. Hale [1969], *Ordinary Differential Equations*. Wiley-Interscience, New York.

J. K. Hale [1977], *Theory of Functional Differential Equations*. Appl. Math. Sci. 3. Springer-Verlag, New York.

J. K. Hale [1988], *Asymptotic Behavior of Dissipative Systems*. Math. Surveys Monogr. 25. Amer. Math. Soc., Providence, RI.

J. K. Hale and W. Huang [1994], Period doubling in singularly perturbed delay equations. *J. Differential Equations* **114**, 1–23.

J. K. Hale and S. M. Verduyn Lunel [1993], *Introduction to Functional Differential Equations*. Appl. Math. Sci. 99. Springer-Verlag, New York.

R. L. Harvey [1991], *Recent advances in neural networks for machine visions*. Proc. of the 3rd Biennial Acoustics, Speech and Signal Processing Mini Conference, IV.1–IV.6. Weston, MA.

R. L. Harvey [1994], *Neural Network Principles*. Prentice-Hall Inc., New Jersey.

S. Haykin [1994], *Neural Networks, A Comprehensive Foundation*. Macmillan College Publishing, New York.

D. O. Hebb [1949], *The Organization of Behavior*. John Wiley & Sons, New York.

A. V. M. Herz [1996] Global analysis of recurrent neural networks. In: *Models of Neural Networks* III (E. Domany, J. L. van Hemmen and K. Schulten, eds.). Springer-Verlag, Berlin, 1–54.

A. V. M. Herz, B. Salzer, R. Kühn and J. L. van Hemmen [1989], Hebbian learning reconsidered: representation of static and dynamic objects in associative neural nets. *Biol. Cybern.* **60**, 457–467.

M. Hirsch[1984], The dynamical systems approach to differential equations. *Bull. Amer. Math. Soc.* **11**, 1–64.

M. Hirsch [1985], Systems of differential equations which are competitive or cooperative II: Convergence almost everywhere. *SIAM J. Math. Anal.* **16**, 432–439.

M. Hirsch [1987], Convergence in neural networks. In: *Proc. of the 1st Int. Conf. on Neural Networks*. San Diego (M. Caudill and C. Butter, eds.), IEEE Service Center, 115–126.

M. Hirsch [1988], Stability and convergence in strongly monotone dynamical systems. *J. Reine Angew. Math.* **383**, 1–53.

M. Hirsch [1989], Convergent activation dynamics in continuous time networks. *Neural Networks* **2**, 331–349.

M. Hirsch and S. Smale [1974], *Differential Equations, Dynamical Systems, and Linear Algebra*. Academic Press, New York.

A. L. Hodgkin and A. F. Huxley [1952], A quantitative description of membrane current and its application to conduction and excitation in nerve. *J. Physiol.* **117**, 500–544.

J. J. Hopfield [1982], Neural networks and physical systems with emergent collective computational abilities. *Proc. Nat. Acad. Sci. U.S.A.* **79**, 2554–2558.

J. J. Hopfield [1984], Neurons with graded response have collective computational properties like those of two-state neurons. *Proc. Nat. Acad. Sci. U.S.A.* **81**, 3088–3092.

F. C. Hoppensteadt [1997], *An Introduction to the Mathematics of Neurons, Modeling in the Frequency Domain* (2nd edition). Cambridge University Press, New York.

L. Huang and J. Wu [1999a], Irregular outstars for learning/recalling dominated spatial patterns. Preprint.

L. Huang and J. Wu [1999b], The role of threshold in preventing delay-induced oscillations of frustrated neural networks with McCulloch–Pitts nonlinearity. Preprint.

L. Huang and Wu [2000], Dynamics of inhibitory artificial neural networks with threshold nonlinearity. To appear in *Fields Inst. Commun.*

L. Huang and Wu [2001], Nonlinear waves in networks of neurons with delayed feedback: pattern function and continuation. Preprint.

D. H. Hubel and T. N. Wiesel [1962], Receptive fields, binocular interaction, and functional architecture in the cat's visual cortex. *J. Physiol.* **160**, 106–154.

D. H. Hubel and T. N. Wiesel [1965], Receptive fields and functional architecture in two non-striate visul areas of the cat. *J. Neurophysiol.* **28**, 229–298.

C. R. Johnson [1988], Combinatorial matrix analysis: an overview. *Linear Algebra Appl.* **107**, 3–10.

E. Kamke [1932], Zur Theorie der Systeme gewöhnlicher Differentialgleichungen, II. *Acta Math.* **58**, 57–85.

D. Kernell [1965], The adaptation and the relation between discharge frequency and current strength of cat lumbosacral motoneurones stimulated by long-lasting injected currents. *Acta Physiol. Scand.* **65**, 65–73.

T. Krisztin and H.-O. Walther [1999], Unique periodic orbits for delayed positive feedback and the global attractor. To appear in *J. Dynam. Differential Equations*.

T.Krisztin, H.-O. Walther and J. Wu [1999], *Shape, Smoothness and Invariant Stratification of an Attracting Set for Delayed Monotone Positive Feedback*. Fields Inst. Monogr. 11. Amer. Math. Soc., Providence, RI.

B. Lani-Wayda and Walther [1995], Chaotic motion generated by delayed negative feedback, part I: a transversality criterion. *Differential Integral Equations* **8**, 1407–1452.

J. P. LaSalle [1976], *The Stability of Dynamical Systems*. Regional Conference Series in Applied Mathematics 25. SIAM, Philadephia.

D. S. Levine [1991], *Introduction to Neural and Cognitive Modelling*. Lawrence Erlbaum Associate, Inc., New Jersey.

R. H. MacArthur [1969], Species packing, or what competition minimizes. *Proc. Nat. Acad. Sci. U.S.A.* **54**, 1369–1375.

M. Mackey and L. Glass [1977], Oscillation and chaos in physiological control systems. *Science* **197**, 287–289.

J. Mallet-Paret [1988], Morse decompositions for differential delay equations. *J. Differential Equations* **72**, 270–315.

J. Mallet-Paret and G. Sell [1996a], Systems of differential delay equations: Floquet multipliers and discrete Lyapunov functions. *J. Differential Equations* **125**, 385–440.

J. Mallet-Paret and G. Sell [1996b], The Poincaré–Bendixson theorem for monotone cyclic feedback systems with delay. *J. Differential Equations* **125**, 441–489.

L. Marcus [1956], Asymptotically autonomous differential systems. In: *Contributions to the Theory of Nonlinear Oscillations*, Vol. **3**, Princeton University Press, Princeton, New Jersey, 17–29.

C. M. Marcus and R. M. Westervelt [1988], Basins of attraction for electronic neural networks. In: *Neural Information Processing Systems* (D. Z. Anderson, ed.), American Institute of Physics, New York, 524–533.

C. M. Marcus and R. M. Westervelt [1989], Stability of analog neural networks with delay. *Phys. Rev. A* **39**, 347–359.

R. M. May and W. J. Leonard [1975], Nonlinear aspects of competition between three species. *SIAM J. Appl. Math.* **29**, 243–253.

W. McCulloch and W. Pitts [1943], A logical calculus of the ideas immanent in nervous activity. *Bull. Math. Biophysics* **5**, 115–133.

E. J. McShane [1944], *Integration*. Princeton University Press, Princeton, New Jersey.

K. Mischaikow, H. L. Smith and H. R. Thieme [1995], Asymptotically autonomous semiflows: chain recurrence and Liapunov functions. *Trans. Amer. Math. Soc.* **347**, 1669–1685.

M. Morita [1993], Associative memory with non-monotone dynamics. *Neural Networks* **6**, 115–126.

V. B. Mountcastle [1957], Modality and topographic properties of single neurons of cat's somatic sensory cortex. *J. Neurophysiol.* **20**, 408–434.

M. Müller [1926], Über das Fundamentaltheoreme in der Theorie der gewöhnlichen Differentialgleichungen. *Math. Z.* **26**, 619–645.

B. Müller and J. Reinhardt [1991], *Neural Networks, An Introduction*. Springer-Verlag, Berlin.

L. Olien and J. Bélair [1997], Bifurcations, stability, and monotonicity properties of a delayed neural network model. *Phys. D* **102**, 349–363.

K. Pakdaman, C. P. Malta, C. Grotta-Ragazzo and J.-F. Vibert [1996], Asymptotic behavior of irreducible excitatory networks of graded-response neurons. Technical Report IFUSPIP-1200, University de São Paulo.

K. Pakdaman, C. P. Grotta-Ragazzo, K. Malta and J.-F. Vibert [1997a], Delay-induced transient oscillations in a two-neuron network. *Resenhas* **3**, 45–54.

K. Pakdaman, C. P. Malta, C. Grotta-Ragazzo and J.-F. Vibert [1997b], Effect of delay on the boundary of the basin of attraction in a self-excited single graded-response neuron. *Neural Comput.* **9**, 319–336.

K. Pakdaman, C. P. Malta, C. Grotta-Ragazzo, O. Arino and J.-F. Vibert [1997c], Transient oscillations in continuous-time excitatory ring neural networks with delay. *Phys. Rev. E* **55**, 3234–3248.

K. Pakdaman, C. Grotta-Ragazzo and C. P. Malta [1998a], Transient regime duration in continuous-time neural networks withd delay. *Phys. Rev. E* **58**, 3623–3627.

K. Pakdaman, C. Grotta-Ragazzo, C. P. Malta, O. Arino and J.-F. Vibert [1998b], Effect of delay on the boundary of the basin of attraction in a system of two neurons. *Neural Networks* **11**, 509–519.

F. J. Pineda [1987], Generalization of back-propagation to recurrent neural networks. *Phys. Rev. Lett.* **59**, 2229–2232.

D. Purves, G. Augustine, D. Fitzpatrick, L. Katz, A. LaMantia and J. McNamara [1997], *Neuroscience*. Sinauer Association, INC., Massachusetts.

W. Rall [1955], Experimental monosynaptic input-output relations in the mammalian spinal cord. *J. Cell. Comp. Physiol.* **46**, 413–437.

P. H. Raven and G. B. Johnson [1995], *Biology* (3rd edition). Wm. C. Brown Publishers, Iowa.

F. Ratcliff [1965], *Mach Bands: Quantitative Studies of Neural Networks in the Retina*. Holden-Day, San Francisco.

C. E. Rosenkilde, R. H. Bauer and J. M. Fuster [1981], Single cell activity in ventral prefrontal cortex of behaving monkeys. *Brain Res.* **209**, 375–394.

T. Roska and L. O. Chua [1992], Cellular neural networks with non-linear and delay-type template elements and non-uniform grids. *Internat. J. Circuit Theory Appl.* **20**, 469–481.

T. Roska and J. Vandewalle [1993], *Cellular Neural Networks*. John Wiley & Sons, New York.

T. C. Ruch, H. D. Patton, J. W. Woodbury and A. L. Towe [1961], *Neurophysiology*. W. B. Saunders, Philadephia.

J. Selgrade [1980], Asymptotic behavior of solutions to single loop positive feedback systems. *J. Differential Equations* **38**, 80–103.

P. Schuster, K. Sigmund and R. Wolff [1979], On ω-limits for competition between three species. *SIAM J. Appl. Math.* **37**, 49–54.

J. Smillie [1984], Competitive and cooperative tridiagonal systems of differential equations. *SIAM J. Math. Anal.* **15**, 530–534.

H. L. Smith [1987], Monotone semiflows generated by functional differential equations. *J. Differential Equations* **66**, 420–442.

H. L. Smith [1988], Systems of ordinary differential equations which generate an order preserving flow, a survey of results. *SIAM Review* **30**, 87–113.

H. L. Smith [1991], Convergent and oscillatory activation dynamics for cascades of neural nets with nearest neighbor competitive or cooperative interactions. *Neural Networks* **4**, 41–46.

H. L. Smith [1995], *Monotone Dynamical Systems: An Introduction to the Theory of Competitive and Cooperative Systems*. Math. Surveys Monogr. 41. Amer. Math. Soc., Providence, RI.

H. Stech [1985], Hopf bifurcation calculations for functional differential equations. *J. Math. Anal. Appl.* **109**, 472–491.

C. Stefanis [1969], Interneuronal mechanisms in the cortex. In: *The Interneuron* (M. Brazier, ed.). University of California Press, Los Angeles, 497–526.

P. van den Driessche and X. F. Zou [1998], Global attractivity in delayed Hopfiled neural network models. *SIAM J. Appl. Math.* **58**, 1878–1890.

H.-O. Walther [1977], *Über Ejektivität und periodische Lösungen bei Funktionaldifferential-gleichungeng mit verteilter Verzögerung*. Habilitationsschrift, Universität München.

H.-O. Walther [1979], On instability, ω-limit sets and periodic solutions of nonlinear autonomous differential delay equations. In: *Functional Differential Equations and Approximation of Fixed Points* (H.-O. Peitgen and H.-O. Walther, eds.). Lecture Notes in Math. 730, Springer-Verlag, New York, 489–503.

H.-O. Walther [1998], The singularities of an attractor of a delay differential equations, *Funct. Differential Equations* **5**, 513–548.

H. R. Wilson and J. D. Cowan [1972], Excitatory and inhibitory interactions in localized populations of model neurons. *Biophys. J.* **12**, 1–24.

J. Wu [1998], Symmetric functional differential equations and neural networks with memory. *Trans. Amer. Math. Soc.* **350**, 4799–4838.

J. Wu [1999], Dynamics of neural networks with delay: attraction and content-addressable memory. In: *Proc. Internat. Conference on Dynamical Systems and Nonlinear Mechanics* (K. Vajravelu, ed.), Florida, 1999.

J. Wu, T. Faria and Y. Huang [1999], Synchronization and stable phase-locking in a network of neurons with memory, *Math. Comput. Modelling* **30**, 117–138.

S. Yoshizawa, M. Morita and S. I. Amari [1993], Capacity of associative memory using a nonmonotonic neural model. *Neural Networks* **6**, 167–176.

Index